Praise for

THE SWEET SPOT

"Bloom writes as if speaking, which brings a welcome immediacy to his explorations. . . . The effect, simultaneously authoritative and chummy, is engaging." —*Harper's Magazine*

"An intriguing scientific investigation into why suffering, from mountaineering to BDSM, so often leads to satisfaction. . . . Bloom has a cheerful writing style that's impossible to dislike." —*Guardian*

"Paul Bloom will change the way you think. Perhaps suffering isn't a bad thing? He explains why the experience of pain enhances subsequent pleasure and that a life without it would actually be boring."
—GoodMorningAmerica.com

"*The Sweet Spot* is lucid and elegantly written throughout so that there's little suffering involved in reading it—in this, it's reminiscent of Michael Sandel and Martha Nussbaum. A bracing, convincing argument that toil, torment, and tribulation can be good things."
—*Kirkus Reviews*

"This book will challenge you to rethink your vision of a good life. With sharp insights and lucid prose, Paul Bloom makes a captivating case that pain and suffering are essential to happiness. It's an exhilarating antidote to toxic positivity." —**Adam Grant, number one *New York Times* bestselling author of *Think Again* and host of the TED podcast *WorkLife***

"Paul Bloom can always be counted on to take your confident assumptions about humanity and turn them upside down. With *The Sweet Spot,* he's done it again! But this time, his investigations into pain and suffering, pleasure and meaning, ask—and answer—the perennial question of what makes life worth living. You won't want to miss this eloquent and erudite book."

—Susan Cain, author of *Quiet*

"Paul Bloom has a gift for spotting paradoxes in human nature and resolving them with deep, satisfying explanations, and this lucid and fascinating book does it again for our puzzling masochisms."

—Steven Pinker, Johnstone Family Professor of Psychology at Harvard University and author of *How the Mind Works* and *Rationality*

"Provocative, fascinating, and insightful—in other words, just what you'd expect from Paul Bloom, one of the world's best writers and deepest thinkers about human behavior. His argument about why we sometimes seek sorrow, fear, and pain is, paradoxically, a pleasure to read. So get out your highlighter and clear your calendar, because once you open this book, you won't be able to put it down."

—Daniel Gilbert, Edgar Pierce Professor of Psychology at Harvard University and author of *New York Times* bestseller *Stumbling on Happiness*

"A laugh-out-loud-funny and totally thought-provoking tour of the most curious parts of human pleasure. With tantalizing examples you can't wait to tell your friends, Bloom provides a fun and theoretically insightful journey into our species' strangest forms of enjoyment. It's a book that will definitely hit your sweet spot!"

—Laurie Santos, Professor of Psychology at Yale University and host of *The Happiness Lab* podcast

"Paul Bloom, one of the best writers we have about the human condition, has done it again. What a fascinating book! *The Sweet Spot* is a profound meditation on happiness, family, and meaning. This book provides enough challenging ideas to give you just a bit of beautiful discomfort, but it is, above all, a joy to read."

—**A. J. Jacobs, author of *It's All Relative***

"Paul Bloom is a phenomenal psychologist. His research is always thought-provoking and his writing clear and eloquent. I eagerly look forward to seeing what he tackles next."

—**Maria Konnikova, author of *The Biggest Bluff***

"This delightful and wonderfully written book manages to ask one of the most significant questions in modern thought: What is happiness, really? Paul Bloom gets to the heart of the matter, taking on a truly important mission: to illustrate how complex and rich human happiness really is. It turns out that, like all things human, happiness can be confusing, paradoxical, embarrassing, and sometimes just bizarre."

—**Greg Lukianoff, president and CEO of the Foundation for Individual Rights in Education and coauthor of *The Coddling of the American Mind***

"Forget your favorite motivational speaker for a moment and read this book. Paul Bloom gives you the real scoop on what it takes to live a good life, and it's not what you may think. Beyond pleasure, beyond joy, and even beyond happiness lies something deeper, more meaningful, and more transcendent. Chosen wisely, suffering can get you there. Paul Bloom shows you how."

—**Scott Barry Kaufman, author of *Transcend* and host of *The Psychology Podcast***

"In 1982 I rode my bicycle three thousand miles in ten days in the Race Across America, resulting in crushing fatigue and almost unimaginable suffering. Until I read Paul Bloom's book on the pleasures of suffering, I never fully grasped why it was one of the most meaningful experiences of my life. Now I understand how struggles embolden one to take on new challenges and better deal with the slings and arrows of outrageous fortune that most of us encounter in our lives, and to find meaning therein."

—Michael Shermer, publisher of *Skeptic* magazine and author of *The Moral Arc*

THE SWEET SPOT

ALSO BY PAUL BLOOM

Against Empathy: The Case for Rational Compassion

Just Babies: The Origins of Good and Evil

How Pleasure Works:
The New Science of Why We Like What We Like

Descartes' Baby:
How the Science of Child Development
Explains What Makes Us Human

How Children Learn the Meanings of Words

THE SWEET SPOT

THE PLEASURES OF SUFFERING AND THE SEARCH FOR MEANING

PAUL BLOOM

An imprint of HarperCollins*Publishers*

HarperCollins books may be purchased for educational, business, or sales promotional use. For information, please email the Special Markets Department at SPsales@harpercollins.com.

A hardcover edition of this book was published in 2021 by Ecco, an imprint of HarperCollins Publishers.

FIRST ECCO PAPERBACK EDITION PUBLISHED 2022

Designed by Paula Russell Szafranski
Illustration on title page © Shutterstock.com

Library of Congress Cataloging-in-Publication Data has been applied for.

ISBN 978-0-06-291057-8 (pbk.)

22 23 24 25 26 LSC 10 9 8 7 6 5 4 3 2 1

For Christina,

sui generis

I knew what it felt like to lie awake, shivering in an inadequate sleeping bag, too cold to sleep and almost too afraid to try. Now, as I slogged through deep snow and deeper darkness toward my tent, tripping and scraping my shins on chunks of broken ice concealed by fresh powder, I reminded myself that I had come here intending to suffer.

—Eva Holland, "Get Schooled in the No-Nonsense Art of Survival," *Outside*

Man, the bravest animal and most prone to suffer, does *not* deny suffering as such: he *wills* it, he even seeks it out, provided he is shown a *meaning* for it, a *purpose* of suffering.

—Friedrich Nietzsche, *On the Genealogy of Morals*

"It hurts just as much as it is worth."

—Zadie Smith, quoting from a condolence letter, in "Joy," *New York Review of Books*

CONTENTS

PREFACE: THE GOOD LIFE

When life is going well, we can forget how vulnerable we are. But reminders are everywhere. There is always the possibility of pain—a sudden ache in the lower back, a cracked shin, the slow emergence of a throbbing headache. Or emotional distress, such as realizing that you just used Reply All to reveal an intimate secret. Such examples just touch the surface. There is seemingly no limit to the misery we can experience, often at the hands of others.

The simplest theory of human nature is that we work like hell to avoid such experiences. We pursue pleasure and comfort; we hope to make it through life unscathed. Suffering and pain are, by their very nature, to be avoided. The tidying guru Marie Kondo became rich and famous by telling people to throw away possessions that don't "spark joy," and many would see such purging as excellent life advice in general.

But this theory is incomplete. Under the right circumstances and in the right doses, physical pain and emotional pain, difficulty and failure and loss, are exactly what we are looking for.

Think about your own favorite type of negative experience. Maybe you go to movies that make you cry, or scream, or gag. Or you might listen to sad songs. You might poke at sores, eat spicy foods, immerse yourself in painfully hot baths. Or climb mountains, run marathons, get punched in the face in gyms and dojos. Psychologists have long known that unpleasant dreams are more frequent than pleasant ones, but even when we daydream—when we have control over where to focus our thoughts—we often turn toward the negative.

Some of this book will explain why we get pleasure from these experiences. It turns out that the right kind of pain can set the stage for enhanced pleasure later on; it's a cost we pay for a greater future reward. Pain can distract us from our anxieties, and even help us transcend the self. Choosing to suffer can serve social goals; it can display how tough we are or, conversely, can serve as a cry for help. Unpleasant emotions such as fear and sadness are part of play and fantasy and can provide certain moral satisfactions. And effort and struggle and difficulty can, in the right contexts, lead to the joys of mastery and flow.

This was going to be the topic of the entire book. It was going to be about how suffering can give rise to pleasure, and was going to be called, not cleverly, *The Pleasures of Suffering.* But as I talked with friends and colleagues and read the work of psychologists, philosophers, and other scholars, I began to have doubts. The theories that work for hot baths and sad songs and spankings turn out not to apply more generally. A lot of the negative experiences we pursue don't provide happiness or positive feelings in any simple sense—but we seek them out anyway. Suffering, yes; pleasure, no.

Think now of a different kind of chosen suffering. People, typically young men, sometimes choose to go to war and, while they don't wish to be maimed or killed, they are hoping to experience challenge, fear, and struggle—to be baptized by fire, to use the clichéd phrase. Some of us choose to have children, and usually we have some sense of how hard it will be; maybe we even know of all the research showing that, moment by moment, the years with young children can be more stressful than any other time of life. (And those who don't know this ahead of time will quickly find out.) And yet we rarely regret our choices. More generally, the projects that are most central to our lives involve suffering and sacrifice. If they were easy, what would be the point?

The importance of suffering is old news. It is part of many religious traditions, including the story in Genesis of how original sin condemned us to a life of struggle. It is central to Buddhist thought—the main focus of the Four Noble Truths. It is at the core of Max Weber's notion of the Protestant work ethic.

Even scholars who disagree on everything else agree about the value of suffering. Toronto, the city where I wrote most of this book, recently hosted a debate between the Canadian psychologist Jordan Peterson, a prominent critic of postmodernism, and Slavoj Žižek, a celebrity philosopher of the far left. Their topic was happiness. In an article about this debate, *The Chronicle of Higher Education* profiled the men, quoting their views and pointing out some similarities. Apparently, they both respect suffering. "The purpose of life," Peterson has written, "is finding the largest burden you can bear and bearing it," while Žižek believes that "the only life of deep satisfaction is a life of eternal struggle." I find these quotes a bit florid—does the struggle really have to be *eternal*?—but still, in their recognition of the centrality of suffering, these men are my brothers.

THE SWEET SPOT does a few things at the same time. Much of it addresses specific questions that I find interesting and I think you will, too. Why do some people like horror movies? Why do some adolescents cut themselves? What's the lure of BDSM? Does unchosen suffering—the death of a child, say—make us more resilient? Does it make us more kind? What will doubling your salary do to your happiness? How will having children influence your sense that your life has meaning?

But *The Sweet Spot* also defends a broader picture of human nature. A lot of people think that humans are natural hedonists, caring only about pleasure. I want to convince you that a close look at our appetite for pain and suffering shows that this view of humanity is mistaken. It turns out that we are inclined toward something deeper and more transcendent.

But I'm not dissing pleasure, either. Instead, this book will defend the idea that there are many things that people want—a view sometimes called motivational pluralism. My view here jibes with that of economist Tyler Cowen, who recently wrote:

> What's good about an individual human life can't be boiled down to any single value. It's not all about beauty or all about justice or all about happiness. Pluralist theories are more plausible, postulating a variety of relevant values, including human well-being, justice, fairness, beauty, the artistic peaks of human achievement, the quality of mercy, and the many different and, indeed, sometimes contrasting kinds of happiness. Life is complicated!

Finally, some of the ideas and findings described in this book can be of practical use. I often think back to two books I read long ago—*Flow*, by Mihaly Csikszentmihalyi, and *Man's Search for Meaning*, by Viktor Frankl. Neither is a self-help book in any simple sense. But each makes claims about human nature and human flourishing that have led many people to rethink how they should live.

I'll talk more about Frankl later, but let me say something about *Flow*. For much of my life I would find myself lost in difficult pursuits, such as training for a marathon or learning to code. But I rarely gave this much thought. Then I read what Csikszentmihalyi had to say about the centrality of such "flow states" to happiness and flourishing. And I realized, for the first time, that these pursuits had value. They were far more important to me than I had realized. And so I consciously decided to try to spend more of my life in the state of flow, which made me happier and more fulfilled.

Such books influenced my life and the lives of many others. I hope *The Sweet Spot* can do the same.

I'VE READ ENOUGH of these types of books to know what's supposed to happen next. I'm now supposed to tell you that we're in a crisis. We are unhappy, adrift, depressed, anxious, lazy, undisciplined, and suicidal. These are the worst of times. What you have here, in these pages, is the solution. This is why you need to read this, why it's so *urgent*.

This is the approach of some excellent books. In *Flow*, Csikszentmihalyi goes on at length about how prosperity has left our lives bereft of meaning—modern Americans, in particular, are miserable. "Genuinely happy individuals are few and far between" he writes. In Emily Esfahani Smith's *The Power of Meaning*, she talks about radical jumps since the 1960s in the number of people suffering from depression, with a corresponding rise in the proportion of antidepressant use, and concludes: "Hopelessness and misery are not simply on the rise; they have become epidemic." In *Lost Connections*, Johann Hari cites the same data and then says that one goal of his book is to explain why there are "so many more people apparently feeling depressed and severely anxious." And just a few pages into his recent bestseller, *The Second Mountain*, David Brooks tells us, "Our society has become a conspiracy against joy," and he goes on to discuss "a shocking rise of mental illness, suicide, and distrust."

Then again, many others argue that these are, relatively speaking, the best of times. The most influential defender of this view is Steven Pinker. In his *Enlightenment Now*, Pinker provides a rich body of data capturing how the world has been getting better. To give just a few examples from the hundreds of case studies he explores, there are increases in life expectancy, food availability, literacy, education, and leisure time. And decreases in child mortality, poverty, war, racism, sexism, and homophobia.

Now, complaints about modern times and belief in progress are compatible. As Pinker is careful to note, "better than it used to be" does not mean "fine." Pinker doesn't deny that the lives of many people are terrible. Further, what he is presenting is the trend so far; it might well be that things are about to get much worse. Perhaps the world is soon to end—due to climate change, say, or nuclear war.

And yet, if you could choose to spend your life at any time in human history, the most rational choice might be right now—and this is particularly the case if you are from one of the poorest parts of the earth, or a woman, or an ethnic minority, or gay, or trans. The fact that millions are escaping extreme poverty each year should cheer us more than it does. If we were better people, it would far outweigh some of the annoyances of modern life that many of us complain about (*People are mean on Twitter! Airline seats are so small!*).

The world has also improved for those of us who are, relatively speaking, doing well. The example I tend to think of is not as impressive as increased life expectancy or the drop in homicide. It is the internet. The machine I'm now tapping away on can access just about any book, movie, or television show, often for free. In seconds, I can be listening to an old comedy album from Steve Martin, rereading a novel by Jane Smiley, or grooving to Alice Cooper. And I'm old enough to remember a time when, if you traveled to another country, it was expensive to even talk to your loved ones back home, and actually being able to see their faces was science fiction. My younger self would have been astonished to see what I did a few weeks ago, sitting in a coffee shop in New Zealand and using my phone to look at and speak with my nieces in Ottawa. If you are unimpressed with all of this, it just shows how quickly we get used to improvements and take them for granted.

You might wonder, though, whether any of this really makes us truly satisfied. Isn't it a great insight that happiness comes from within? "There is nothing either good or bad but thinking makes it so," wrote Shakespeare. Surely we can be miserable in a world of abundance and joyous in the worst of circumstances.

Yes, and there will be a lot of discussion of this in what follows. But it's an even greater truth—one so obvious that hardly anyone talks about it—that the good life is much easier to achieve if you are in physical and emotional comfort. It's hard to be joyous and satisfied if your children are starving to death or you are in agony because of an untreated illness. It would be bizarre if improvements in our life conditions made no difference in our happiness.

And, actually, Pinker notes that there is a positive trend in happiness over time, at least over recent history. In countries where there are data from multiple surveys, people tend to be happier in the most recent one. And most people alive describe themselves as happy. To put a number on it, the World Values Survey found that 86 percent of people they assessed globally describe themselves as "rather happy" or "very happy." When experts insist that their societies are awash in misery, they are unknowingly illustrating one of the big findings in happiness research, which is that people underestimate how happy other people are—we tend to think of ourselves as lucky exceptions.

This good fortune is not evenly distributed, though. Some countries are happier than others. Now, you might be skeptical about how happiness is measured, and we'll talk soon about the vagueness of the word "happy"—among other problems, the word has different meanings in different languages, making comparison difficult. But the precise wording of the question doesn't seem to make a difference in how the results turn out. While some studies, like the World Values Survey, ask about "happiness," others use different methods, such as asking people to rank their lives on a score from 0 (the worst possible life) to 10 (the best possible life).

And, no matter how you slice it, the happiest countries turn out to be just those you would expect, such as the Nordic nations of Norway, Iceland, Finland, Denmark, and Sweden, as well as Switzerland, the Netherlands, Canada, New Zealand, and Australia. They are high in income, with good life expectancy and strong social support. The citizens of these societies report high levels of freedom, trust, and generosity.

These country comparisons tell us some interesting things about the best conditions for human flourishing. As the psychologist Edward Diener and his colleagues point out, both liberals and conservatives have something to crow about. Liberal policies, such as progressive taxation and a strong welfare state, predict happiness. But so do the factors that conservatives emphasize, such as some degree of economic competition (communist countries do poorly on happiness). Other work suggests that, at the individual level, traditional pursuits such as religion and marriage and stable family ties also predict happiness—though, as we'll see, the effects of having children are quite a bit more complicated.

These findings also show that happiness is not fixed. While there are genetic influences on happiness, you can move your happiness up and down by way of your choice of where to live. Miserable? Pack your bags and move to Toronto or Stockholm. Want more misery in your life? Well, there are plenty of countries on the bottom of the list that would be happy to have you. You might object that it's not actually living in the country that influences one's happiness; you might think that Swedes, say, are happy because of Swedish genes or Swedish upbringing and that they would be just as happy if you moved them to Angola or Cuba, two of the saddest places on earth. But this isn't so. Several studies show that while your nation of origin does have some influence, immigrants and native-born citizens of a country tend to be roughly as happy as one another. So, yes, the society in which you live really does influence your happiness.

IF THINGS ARE going so well, why—besides a general intellectual interest—should you worry about the best conditions for a good life?

Well, perhaps you are American. If one does want to make a case for a crisis, the United States is a good place to start. For a country of such wealth, it does relatively poorly, though it ranks high overall (ranking eighteenth out of 156 countries in the most recent World Happiness Report).

More than this, the United States is going through a difficult time. Although some of the claims about this are controversial—for instance, it's unclear whether there really is a loneliness epidemic—there is plainly something wrong here. There has been a great decline in the rate of suicide worldwide (38 percent since the mid-1990s), but the trend in the United States is the opposite—American suicides have shot upward by about 30 percent since 2000. David Brooks describes the situation as "horrific" and notes as well the "slow-motion suicide" seen in the opioid epidemic. He points out that the average life span for Americans has been declining over the past few years, a stunning trend for an affluent society—the last time this happened in the United States over such a long span of time, he notes, was from 1915 through 1918, during World War I and the Spanish flu epidemic, which killed more than half a million Americans.*

Brooks and others see the core problem here as a crisis of meaning, associated with the decline of religious faith, loss of overall purpose, and alienation from flesh-and-blood communities. Describing this crisis, Johann Hari says we have "Facebook friends in place of neighbors, video games in place of meaningful work, status updates in place of status in the world."

---

* As I write this—in August 2020—we are in the midst of another global pandemic. The long-term effects of COVID-19 on our happiness and flourishing is very much an open question.

Such problems have been around since long before social media. In his book *Tribe*, Sebastian Junger describes America at the end of the eighteenth century, with two civilizations fighting over the same land. It was a time, he says, when "factories were being built in Chicago and slums were taking root in New York while Indians fought with spears and tomahawks a thousand miles away." In the course of this conflict, some of the colonists would be kidnapped, usually women and children. Surprisingly, despite the considerable deprivations and the estrangement from family and friends, many of those who were captured liked their new lives. They would marry their captors, become part of their families, and sometimes fight alongside them, sometimes hiding from their rescuers. In some cases, they had to be tied up in order to complete arranged prisoner exchanges, and when they were brought back to their original homes, they often escaped and tried to return to their Native American communities.

It never went the other way around. Benjamin Franklin marveled about this in a letter to a friend in 1753: "When an Indian child has been brought up among us, taught our language and habituated to our customs, yet if he goes to see his relations and make one Indian ramble with them, there is no persuading him ever to return."

Junger asks what the Indigenous communities had that the seemingly more advanced Europeans lacked. His answer is that, for the first time, the captured colonists had a taste of a life filled with meaning and purpose and community.

WE'VE SEEN THAT reasonable people are worried about a lack of meaning in their lives. Some not-so-reasonable people worry about it, too. In *The Unabomber Manifesto*—35,000 words by the domestic terrorist Theodore Kaczynski, who murdered three people and injured many more—Kaczynski distinguishes between three types of goals. There are those that can be satisfied with minimal effort, those that require serious effort, and those that can't be satisfied at all. Kaczynski complains that the middle category has been lost. As Peter Thiel puts it, summarizing the argument: "What you can do, even a child can do; what you can't do, even Einstein couldn't have done." The solution, according to Kaczynski, is to blow up technology and start anew.

Thiel goes on to suggest that this sort of pessimism is common in fundamentalist movements, which offer nothing between the easily known and the unknowable. He notes as well that this attitude is sometimes manifested not in violence but in lassitude, and cites the example of the hipster movement: "Faux vintage photography, the handlebar mustache, and vinyl record players all hark back to an earlier time when people were still optimistic about the future. If everything worth doing has already been done, you may as well feign an allergy to achievement and become a barista."

I'm agnostic myself as to whether or not the modern world suffers from a lack of meaning or purpose, relative to the past. But I do know that many people have something missing in their lives, and that meaningful projects—which will come with pain, difficulty, and struggle—can be the cure for what ails us. This tweet, from Greta Thunberg, one of the best-known activists of our time, captures a pretty typical reaction to finding meaning in one's life:

> Before I started school striking I had no energy, no friends and I didn't speak to anyone. I just sat alone at

home, with an eating disorder. All of that is gone now, since I have found a meaning, in a world that sometimes seems shallow and meaningless to so many people.

Viktor Frankl came to a similar conclusion. In his early years as a psychiatrist in Vienna, in the 1930s, Frankl studied depression and suicide. During that period, the Nazis rose to power, and they took over Austria in 1938. Not willing to abandon his patients or his elderly parents, Frankl chose to stay, and he was one of the millions of Jews who ended up in a concentration camp—first at Auschwitz, then Dachau. Ever the scholar, Frankl studied his fellow prisoners, wondering about what distinguishes those who maintain a positive attitude from those who cannot bear it, losing all motivation and often killing themselves.

He concluded that the answer is meaning. Those who had the best chance of survival were those whose lives had broader purpose, who had some goal or project or relationship, some reason to live. As he later wrote (paraphrasing Nietzsche), "Those who have a 'why' to live, can bear with almost any 'how.'"

As a psychiatrist, Frankl was interested in mental health. But his plea for a life of meaning—a central part of the therapy he developed once he left the camps—wasn't merely based on the notion that this would provide happiness or psychological resilience. He believed that this is the sort of existence we should want to pursue. He was sensitive to the distinction between happiness and what Aristotle described as eudaemonia—literally "good spirit," but actually referring to flourishing in a more general sense. It was eudaemonia that mattered to Frankl.

When the war ended and he was released from the camps at the age of forty, Frankl himself had nothing. His wife and mother and brother had been murdered by the Nazis. And so he rebuilt. He started to work again as a psychiatrist. He remarried. He had children, and then grandchildren. He wrote, starting with the classic Holocaust narrative *Man's Search for Meaning.* When Frankl died, at the age of ninety-two, he had just completed a final book. It was a rich life, replete with both meaning and pleasure.

I WANT TO be clear about my position here. I'm *not* going to argue that people who are miserable need more suffering. To tell someone who is on the verge of suicide that they need more pain in their life would be cruel if it weren't so ridiculous.

In fact, when it comes to certain sorts of suffering, I'm more skeptical than many. We'll see later in this book that there are a lot of researchers who say that bad experiences in life are actually good for you—these researchers speak about post-traumatic growth, an increase in kindness and altruism, increased meaning in life. I don't buy any of this. Unchosen suffering is awful; avoid it if you can.

So what *am* I going to argue? This book defends three related ideas. First, certain types of chosen suffering—including those that involve pain, fear, and sadness—can be sources of pleasure. Second, a life well lived is more than a life of pleasure; it involves, among other things, moral goodness and meaningful pursuits. And third, some forms of suffering, involving struggle and difficulty, are essential parts of achieving these higher goals, and for living a complete and fulfilling life.

I'LL END THIS preface with a confession. Before delving into these topics, I had some exposure to what people were saying about happiness, and my impression was not positive. I was pretty scornful. I felt a lot of happiness research was superficial, with unsupported claims and bad philosophy. I thought of it as more scam than science.

Part of the reason for my bad attitude was that, like a lot of us, I was getting these ideas filtered through sources like TED talks and self-help books. This has a distorting effect. If you want to get onstage and stay there, if you want money and fame, you are wise to offer solutions to life's problems, regardless of how strong the scientific data are. I don't want to exaggerate this. There are popular representatives of every field who are honest dealers. But there are also hucksters, and the happiness business has a lot of them.

Years ago, I was one of a group of invited speakers at a small meeting of very wealthy people in Florida. The night the conference began, the hosts brought out a surprise speaker after dinner. The man wasn't just an academic or a businessman like the rest of us; he was *famous*—we all burst into applause when we heard his name. I don't want to single him out, so I'll just say that he is one of the most famous motivational speakers alive. I knew of the man's reputation, so I was excited to hear what he had to say.

As our hosts advertised, this was a transformative experience. Just not in the way they had envisioned. Soaked in sweat, he told us about some supposedly life-changing experimental findings from psychology—most of them false and long discredited. He took comedy bits from HBO specials and passed them off as his own. And he was generally incoherent, at one moment spouting homilies about infinite love and then, a minute later, getting us to work with him on a David Mamet–like exercise where he asked us to turn to the person sitting closest to us and scream, *"I own you!"* I tried to be a good sport and take it seriously, but I was sitting next to a historian, and when I screamed at her, she couldn't stop giggling.

Much of what we are told about happiness and the good life should not be trusted. But I now see my earlier view as uncharitable and ultimately mistaken. I no longer feel that if I had a long weekend, I could clean the whole field up. There's a sophisticated core to the science of well-being, which includes both self-defined "positive psychologists" and scholars who wouldn't be caught dead being associated with that group. There is careful empirical work and deep theorizing. Some scholars I've been influenced by include (and this is a partial list): Mihaly Csikszentmihalyi, David DeSteno, Edward Diener, Daniel Gilbert, Jonathan Haidt, Daniel Kahneman, Sonja Lyubomirsky, and the founder of the field of positive psychology, Martin Seligman. *The Sweet Spot* has also been influenced by excellent books by Emily Esfahani Smith and Brock Bastian that explore similar themes—Smith on meaning and Bastian on pain and suffering.*

But this book is more than a summary of others' ideas and research. The topics I'll be addressing, concerning the pleasure we get from pain and the centrality of suffering in our lives, have been underexplored, and this book goes off in some unusual directions. Some ideas and arguments I make are firmly grounded in scientific research; others are more tentative, and I'll try to be clear on which is which.

---

* I'll add that this is a fast-moving field, and any book is going to be at least a little bit out of date by the time it's published. The best recommendation I have for keeping up with the latest discussions is the podcast *The Happiness Lab*, by my friend and colleague Laurie Santos, at https://www.happinesslab.fm.

Also, as Walker Percy once wrote, "Fiction doesn't tell us something we don't know, it tells us something we know but don't know that we know." Sometimes this is true for psychology as well. I am going to tell you things you didn't know you knew.

1

SUFFER

My younger son has an appetite for pain. He was always the kid who would get into a slapping competition with his buddies or challenge you to a wasabi-eating contest. For his senior project in high school, he climbed Everest. Not the real Everest—he did have to go to class, after all—but Everest is 29,029 feet tall, so Zach went to the climbing gym late each afternoon and climbed up and down for hours (just under a thousand feet a day, four to five days each week, for thirty days), keeping a blog of where he would be and what he would be seeing if he were actually departing from the base camp in Nepal, ascending the mountain, and returning. It was unpleasant and grueling, and he complained bitterly, and he loved it.

I bet you've done something similar. Perhaps you camp, forsaking your soft bed and hot shower. Or you might cycle, a sport where elite competitors rhapsodize about "sweet pain," described by one cyclist as "that breathless and bone-tired feeling followed by a crooked smile when you see your time . . . the headspace you go to ignore your screaming legs approaching the crest of [a] really long brutal climb on the bike when all you do is pray to the endorphin gods."

I'm no jock, but long ago I ran the New York Marathon. When I decided to do this, I was terribly out of shape, so it involved more than a year of preparation, some of it during a cold New England winter. I remember what it was like to run in the morning darkness, to feel my face go numb, to nurse my blisters and muscle aches. But these are memories I cherish.

Then there are the more passive masochistic pleasures. There's no real mystery as to why we enjoy feeling exhilaration and awe, or vicarious triumph, or sexual desire. But what's going on with horror? Years ago, I came across my older son doing his physics homework while watching, on his laptop, an artsy French cannibal movie called *Raw*. I took one look and it ruined my afternoon. This sort of thing is too much for me; when I talk about torture porn and the like, I'll be going off secondhand descriptions.

(The same son cheerfully introduced me to a Reddit forum called r/wince, which is exactly what it sounds like. I clicked on it just now and the first item for today is a photo titled "Staple through the finger. Went through bone." I went weak at the very thought and immediately moved to another screen.)

Maybe you like this sort of thing—maybe you've put down the book and are now checking out Netflix or Reddit to get a good look at what I'm talking about—or maybe you're more like me, a sensitive soul. But still, everyone has some taste for aversive experiences. My own viewing pleasures include television series like *The Sopranos* and *Breaking Bad* and *Game of*

Thrones. These are all violent—with rape, murder, and torture—and include depictions of all sorts of suffering and loss. But they have their hold on me, and I bet that something similar, perhaps with melancholy instead of violence, pushes your buttons.

The specific sort of suffering we like to indulge in, and how severe we want this suffering to be, differs from person to person. I like spicy curries and roller-coaster rides myself. Hot baths? Yes, though not too hot. Distance running? Yes. BDSM? None of your business. But while there are interesting differences, nobody is immune to the lure of suffering.

BEFORE GOING FURTHER, I want to address a semantic concern. I'll be using the words "pleasure" and "pain" in just the way everyone else uses them—roughly, to refer to the experiences that, respectively, make you go *Ahh!* or *Ouch!* But I am also going to talk about negative experiences that aren't physically painful, such as working long hours on a difficult project, obsessing over sad memories, or choosing to go without food when you're hungry. Sometimes I'll call this "suffering." This fits the standard dictionary definition: the state of undergoing pain, distress, or hardship. (This definition doesn't say that it has to be *a lot* of pain, distress, or hardship.)

But I've come to realize that there are some who are uncomfortable with, even offended by, this word choice. I once described some mundane activity (mild electrical shocks in the lab) as suffering and had an older woman angrily tell me that her parents had gone through horrific experiences during World War II, and *that* was suffering. For her, my broad usage minimized what happened to them. I get it. I feel the same way when I hear someone describe an experience such as waiting in a long security line at the airport as "torture." This might be fine as comic exaggeration, but to say it seriously is offensive—it trivializes the real thing.

I wish that English had a richer vocabulary here, so we could more easily make the distinctions. But it doesn't, so I'll continue to use the word "suffering" for the full continuum of negative experience. Just as pressing a sore tooth with your tongue is still pain, forms of mild suffering are still suffering. But if you don't like this way of talking, just mentally translate my usage of this word to the more awkward but maybe more accurate "experiences that are typically aversive, for either physical or psychological reasons," and we'll be on the same page.

THIS BOOK WILL explore two different sorts of chosen suffering. The first involves spicy food, hot baths, frightening movies, rough sex, intense exercise, and the like. We'll see that such experiences can give pleasure. They can increase the joy of future experiences, provide an escape from consciousness, satisfy curiosity, and enhance social status. The second is the sort involved in climbing mountains and having children. Such activities are effortful and often unpleasant. But they are part of a life well lived.

These two sorts of chosen pain and suffering—for pleasure and for meaning—differ in many ways. The discomfort of hot baths and BDSM and spicy curries is actively pursued; we look forward to it—the activity wouldn't be complete without it. The other form of suffering isn't quite like that. When training for a marathon, nobody courts injury and disappointment. And yet the possibility of failure has to exist. When you start a game, you don't want to lose, but if you know you will win every time, you're never going to have any fun. So, too, with life more generally.

The impossibility of failure is one of the weaknesses of daydreaming. The behavioral economist and psychiatrist George Ainslie once complained that daydreams suffer from a "shortage of scarcity." We can choose to put ourselves into a bind, but we can also choose to get out of it. This freedom can strip away much of the pleasure we get from solitary fantasy.

This is why, in case you were wondering, omnipotence is boring. If there were no kryptonite, who would care about Superman's adventures? Actually, true omnipotence would be misery. There is an old *Twilight Zone* episode that elaborates on this point. A gangster dies and, to his surprise, wakes up in what seems to be paradise. He gets whatever he wants—sex, money, power. But boredom sets in, and then frustration, and finally he tells his guide that he doesn't belong in heaven. "I want to go to the other place," he says. And his guide responds that this isn't heaven; he is already in the other place.

PHRASES LIKE "pain that's also pleasure" and "the joy we can get from suffering" make sense. Examples like saunas and torture porn make clear that we appreciate the lure of certain forms of pain and distress. "Hurts so good," says the philosopher-songwriter John Mellencamp, and we nod along. But if you think about it, the idea is a bit strange, even paradoxical.

After all, badness seems to be part of the very notion of pain. In a classic paper, the philosopher David Lewis imagines a madman who feels pain that's different from ours. While our pain makes us wish for it to stop and might make us yell or cry, his pain makes him act in strange ways—he thinks about mathematics; he crosses his legs and snaps his fingers. And the madman that Lewis imagines has no motivation to avoid pain or make it go away once he has it.

Lewis's analysis here is subtle, but my reaction, and possibly yours, is that it's not really pain after all. The madman might call it pain, but this confusion just reflects his mental illness. It can't be pain if it has no association with the negative, so he's wrong to describe it as such.

And this is why pleasure from pain is so puzzling. Consider two definitions that pop up when you type the words into a search engine.

> Pleasure: a feeling of happy satisfaction and enjoyment
> Pain: a highly unpleasant physical sensation caused by illness or injury

These look like opposites. If you check out the more technical definition by the International Association for the Study of Pain Task Force on Taxonomy, you'll see that pain is "an unpleasant sensory and emotional experience arising from actual or potential tissue damage or described in terms of such damage"—and there it is again, the word "unpleasant." How can an experience be pleasant and unpleasant simultaneously?

According to a certain way of seeing things, it can't. Suppose every moment of experience corresponds to a number on a scale from 0 to 10, with low numbers being awful states that you avoid and high numbers being positive states you pursue. You can't have a state with both a low number and a high number associated with it. It would be like pouring a bath that's both hot and cold. Impossible—it can be hot, cold, or in between; it can be hot at 8 p.m. and cold at 8:15 p.m.; it can even be hot on the right side and cold on the left side. But the same water just can't be simultaneously hot and cold.

To see the puzzle in a different way, think about the function of these psychological states. Jeremy Bentham said that "nature has placed mankind under the governance of two sovereign masters, pain and pleasure," and he saw them as inherently opposing forces, pushing us in different ways: approach and avoidance, carrot and stick. But how can you approach and avoid at the same time?

We'll talk about Freud in a little while, but I'll just note here that whatever one might say about his views, he did appreciate the weirdness of the phenomenon. He writes that since the primary aim of a person "is the avoidance of unpleasure and the obtaining of pleasure," it follows that seeking out pain is "incomprehensible." In such cases, "it is as though the watchman over our mental life were put out of action by a drug."

PERHAPS THE WAY out of this puzzle is to conclude that pain is never pleasurable. We seek out pain, sure, but maybe we do so only because this provides other benefits. This sort of trade-off is the stuff of life. You run outside on a chilly day, shivering and uncomfortable, to retrieve an important package that has been left up the walk. Or undergo a painful operation to fix a long-standing medical condition. Or sit, bored and unhappy, in a government office in order to renew your driver's license. Or even withstand torture so as not to reveal the identities of your comrades. There are many reasons to choose pain and suffering that don't deny their awfulness. And the next chapter, on benign masochism, includes a lot about how we choose pain to obtain pleasure just a few seconds afterward. Such explanations don't deny the badness of pain.

But it turns out that pain itself need not be negative. We can get some hint of the complexity here by looking at certain clinical conditions.

You may have heard of congenital analgesia. People who suffer from this can feel themselves being cut or hit, but they don't register these experiences as pain, and so have no intrinsic motivation to avoid them. Most people with this condition don't live past their twenties, and this illustrates the importance of pain, both in preventing injury and allowing injuries to heal.

A more puzzling syndrome is pain asymbolia. This is a condition wherein people feel pain and describe their experience as painful—but they don't find the pain to be unpleasant. They offer up parts of their body to doctors and scientists for intrusions that, for you or me, would be agonizing. But it's not as if they are numb; one patient reported, "I feel it indeed; it hurts a bit, but it doesn't bother me; this is nothing." This disorder is associated with damage to parts of the brain such as the posterior insula and the parietal operculum, areas that, more generally, respond to threat. Such a syndrome should open our eyes to the idea that the experience of pain need not be inherently unwelcome.

These two sorts of pain syndromes—congenital analgesia and pain asymbolia—correspond to a distinction sometimes made between two kinds of analgesics. There is the usual kind, which dulls or obliterates pain, and then there's another sort (and morphine is sometimes described in this way) that, while it does have a powerful dulling effect, gives you a sort of pain asymbolia. You feel the pain, but it bothers you less.

The philosopher Nikola Grahek notes that we can get a glimmer of what pain asymbolia feels like in our everyday lives. He asks his reader to imagine going to the doctor because of a dull, nagging pain in the upper-left chest, radiating down the arm. You are worried that it's a heart attack, but the doctor reassures you that it's muscle inflammation and will fade soon. Your fear will go away and "you will take a carefree attitude toward the pain, although the pain will still be there and will still be felt as unpleasant."

Sometimes the changing reaction to pain comes from a change of attitude. The writer Andrea Long Chu talks about a long, painful preparation for the surgery that would transform her penis into a vagina, and she begins by describing pain as it is often felt: "All bodily pain begins with shock at the audacity of physical trespass." But she then notes that over the months, "we reached a cautious détente, the pain and I, acknowledging each other's presence on the tacit condition of mutual noninterference, like exes swapping nods at a holiday party."

This is said to be one of the powers that come from meditative practice. Robert Wright talks about an experiment he tried during a meditation retreat:

> A tooth—which turned out to require a root canal—had started hurting me whenever I drank anything. The pain was sharp and could be excruciating, even if what I was drinking was at room temperature. So, just to see what would happen, I sat down in my room and meditated for thirty minutes and then took a giant swig of water and made a point of bathing the tooth in it.
>
> The result was dramatic and strange. I felt a throbbing so powerful that I got absorbed in its waves, but

the throbbing didn't consistently feel bad; it was right on the cusp between bitter and sweet and just teetered between the two. At times it was even awesome in the old-fashioned sense of actually inspiring awe—breathtaking in its power and, you might even say, its grandeur and its beauty. Maybe the simplest way to describe the difference between this and my ordinary experience with tooth pain is that there was less "youch!" than usual and more "whoa!" than usual.

SUCH CASES SUGGEST that pain need not be bad. But both scientific research and everyday experience point to an even stronger claim: Pain can be good. The picture we presented earlier, of a scale from 0 to 10, is wrong. Perhaps other creatures work this way, with pain and pleasure on a single continuum. But for people, something can be both a 0 and a 10. Negative experiences and positive experiences—pain and pleasure—are not opposites; thinking of them like low temperatures and high temperatures is a mistake.

How is this possible? The answer lies in the human capacity to interpret and respond to experiences. We can be made to feel happy, sad, angry, ashamed, or amused by events in the world, but we can also be made to feel happy, sad, angry, ashamed, or amused by our *responses* to events in the world. (And, sometimes, happy, sad, angry, ashamed, or amused by our responses to our responses to events in the world, but for the sake of simplicity, let's put this aside.)

Take fear as an example. You are charged by a tiger and you are terrified. This fear is an adaptive response we share with other creatures. Adrenaline is released, heart rate increases, blood flows into the muscles, and the digestive system slows or shuts down as the body prepares to fight or flee. (Eminem summed this up nicely in his description of someone facing a high-risk, high-reward social competition: "His palms are sweaty, knees weak, arms are heavy / There's vomit on his sweater already.") Goose bumps might arise, an atavistic response betraying our ancestry as hairy creatures. There is an increase in arousal and a focusing of attention. One thing about fear: it's not boring.

This sort of experience is usually negative. Being charged by a tiger is the worst. But the badness of the experience isn't a result of the fear. It is because it would be awful to be maimed or killed by a tiger. Suppose you know that there's no real risk (perhaps you are in a virtual reality simulation). You might still experience fear—your body might react in much the same way—but it's not necessarily bad fear. It might be fun fear.

People pay for this sort of experience, after all. Haunted houses and scary movies are very popular. And we know that the fear is part of the appeal here. In studies I'll describe later, researchers found that fans of horror movies experienced just as much fear when they watched a film like *The Exorcist* as those who dislike horror movies. Contrary to some theories, then, those who enjoy frightening movies are not emotionally numb. Rather, they like the fear. In fact, the more fear they experience, the more pleasure they get.

To take another example, anger is usually a response to perceived injustice, and so angry experiences are often negative ones. But you can savor being angry, perhaps fantasizing about revenge or enjoying the sensation of righteous outrage. And anger can be useful. In a clever study, Maya Tamir and Brett Ford found that when people were motivated to confront (as opposed to cooperate with) a negotiation partner, they were likely to try to increase their anger, and they expected this anger to help. And they were right to do so: angry negotiators were more successful in their negotiations.

Or take sadness. This is usually a response to negative events, but there's a pleasure to be found in sulking, wallowing in our misery, so long as the bad thing isn't that bad. (Nobody enjoys the feeling of grief after the death of someone they love.) In another study, people were shown sad movies, and the extent of their sadness predicted how much they wanted to continue watching. The sadness didn't correspond to something that was actually bad, so it was, or at least had the potential of being, pleasant sadness.

Also, we do seem to get something out of songs and compositions that are engineered to evoke sadness, such as the work of Lana Del Rey and Adele, or classical pieces like *Adagio for Strings* or the requiems of Mozart and Verdi. Studies find that while listening to such classical compositions, people appreciate the sadness they convey and take pleasure from it, claiming to feel, among other things, tenderness and nostalgia.

Why do sad songs have this appeal? Maybe we just enjoy experiencing sadness in a safe context, savoring the feeling without any real-world concerns. Maybe there are more specific rewards. In an essay for one of my undergraduate seminars, Emily Cornett wondered why someone whose heart has recently been broken might enjoy listening to a breakup song. She suggests that the song reassures the person that they're not alone, that there exist others who have felt just the same. Cornett also notes the importance here, as in all negative experiences, of choice. To have just had a breakup and then suddenly, by chance, to hear a song like Adele's "Someone Like You" would likely be an unpleasant experience. We like to have some control over when we burst into tears.

Just about any emotion can be transformed in this way. In the movie *The Big Short*, Mark Baum, played by the actor Steve Carell, is portrayed as perpetually enraged. When he says that his wife told him his job is making him unhappy, one of his colleagues responds, "But you're happy when you're unhappy," and Baum agrees.

We've been talking about how negative experiences can be a source of pleasure. But does the opposite happen? Can you reappraise positive experiences so that they become negative? Apparently, yes. Some people with depression become unwilling to experience positive emotions. They might believe, for instance, that they don't deserve to be happy or that a happy experience now will only set the stage for some bitter unpleasantness in the future. Along with pleasant sadness, then, you might have miserable joy.

There are also cross-cultural differences. Studies find that East Asians are more skeptical of happiness than Westerners. Arguably, Asian cultures have a more "dialectical" appreciation of happiness and sadness, nicely illustrated by this quote from the *Tao Te Ching*:

> Happiness rests in misery
> Misery hides in happiness
> Who knows where they end

You don't have to be Taoist to appreciate the mixed nature of the emotions. One recent study tested subjects from the United States, Canada, China, and South Korea and asked about the six emotions that psychologists see as the most universal and foundational: sadness, fear, disgust, anger, happiness, and surprise. For each, the subjects were asked how positive and how negative the emotions seemed to them. As you would expect, people tended to see sadness, fear, disgust, and anger as mostly negative and happiness and surprise as mostly positive. But though there were cultural differences, it turns out that the judgments of all of the emotions were mixed, with quite a bit of positivity for sadness and even a smidgen of negativity for happiness.

Let's again consider actual physical suffering. We discussed the example of the seeming heart attack, but now imagine yourself at the end of a marathon, and you're not in the best shape. Your heart is pounding and you are soaked with sweat and panting for air. If you suddenly felt that way while sitting on the bus or trying to go to sleep, it would be one of the worst moments of your life; you would think you were going to die. But in the context of the marathon, where it makes sense and reflects your hard work, this seemingly aversive experience can be part and parcel of a great accomplishment, something to savor.

Or take getting punched in the face. This might seem like a particularly bad experience, but it doesn't have to be, at least not entirely. In Josh Rosenblatt's narrative about his evolution as a mixed martial arts fighter, he says that the first time you get hit in the face, you are overwhelmed with fear. Then you reach a second stage where you react with anger and shame. But after that:

> You start to *love* getting hit in the face, and then you start to *need* getting hit in the face. You court danger now; life starts to feel empty without it. . . . It pumps the blood faster through your veins; it makes your eyes water and

your heart race. It makes the world shimmer. It reminds you of your mortality even as it snaps you into that concentrated present moment mystics call eternity.

I'm willing to take Rosenblatt's word on all this (in my own martial arts experience, I never got past anger and shame), but surely the context matters. If Rosenblatt was standing in line for the movies and someone popped him one, I bet it would not be life-validating. If the world shimmered, it wouldn't be in a good way. But still, in the proper situation, he is right—the terrible can morph into the transcendent.

We've been domesticating the desire for pain and suffering, making it less mysterious, and we'll continue to do so for the next few chapters. If you believe that humans seek happiness, chosen pain is no longer an obvious counterexample.

But is happiness really what we want? Many people think so. Freud writes that when it comes to people's primary motivation, "the answer to this can hardly be in doubt. They strive after happiness; they want to become happy and to remain so. This endeavor has two sides, a positive and a negative aim. It aims, on the one hand, at an absence of pain and displeasure, and, on the other, at the experiencing of strong feelings of pleasure." Blaise Pascal was even blunter: "All men seek happiness. This is without exception." And, to make clear how serious he is, he later adds: "This is the motive of every action of every man, even of those who hang themselves."

These quotes are from Daniel Gilbert's excellent book *Stumbling on Happiness*, and they summarize Gilbert's own view. He thinks that we all strive for happiness and that this is a perfectly good and rational pursuit. Gilbert is aware that some philosophers push back on this, but he thinks that they just have a too narrow understanding of what it is to be happy. As he puts it, many philosophers see the desire for happiness as akin to the desire for a bowel movement, "something we all have, but not something of which we should be especially proud." Less graphically, they see happiness as bovine contentment, reflecting a sort of dullness.

But you should reject this analysis, Gilbert argues, and instead see happiness as a certain feeling that can be sparked by all sorts of experiences—it can be low, but it can also be high.

Ursula K. Le Guin makes a similar point in her brief story about the land of Omelas, where people live the most wonderful lives—though this comes at a terrible price. (If you haven't read this, you should; the link is in the notes.) After telling us how happy the citizens of Omelas are, she cautions us not to jump to the conclusion that they are simple, bland, or unintelligent, and she adds, "The trouble is that we have a bad habit, encouraged by pedants and sophisticates, of considering happiness as something rather stupid. Only pain is intellectual, only evil interesting."

I think this is all fair enough. But these qualifications also illustrate what's wrong with saying "People want to be happy." The problem isn't that this claim is mistaken. The problem is that it's too vague to be helpful.

I'm far from the first to voice this concern. Many researchers in positive psychology avoid talking about happiness, replacing it with phrases like "subjective well-being." One reason to avoid the term is that researchers often want to make comparisons between countries, and the words "happiness" and "happy" don't translate well. An English speaker can say, "She is happy sitting here reading," while speakers of French and German can't use the equivalent words *heureux* and *glücklick* in the same context. That is, the English word is more expansive than its equivalents in other languages. It's easier to be "happy" if you speak English. (Which does not mean, of course, that it's easier to be happy.)

A further issue is that some people distinguish happiness from morality, and some don't. When Freud talks about "strong feelings of pleasure," he doesn't mention improving people's lives or making the world a better place. But others see happiness as morally laden. The philosopher Philippa Foot gives the example of a Nazi commandant who experiences pleasant mental states—Foot argues that he can't be truly happy, because he is not living a good life. For her, happiness requires goodness.

Maybe you don't share Foot's intuition—I don't; I can easily imagine a happy Hitler—but there are some experiments suggesting that our feelings about the goodness of someone's life do have some influence on whether we see them as happy. Some experimental philosophers (including my former student Jonathan Phillips and my colleague Joshua Knobe) did a series of studies where they told subjects about two people who had the same positive mental states. They found that the subjects were more likely to think that a person is happy if she is living a morally virtuous life as opposed to a selfish, hedonistic one. Foot is on the right track, then: happiness—in at least one sense of the term—is related to morality.

THE MORE GENERAL problem with the claim that we want to be happy is that happiness can be used to refer to (at least) two different things. A question like "How happy are you?" can refer to your experience right now ("I am very happy: I am eating M&Ms!") or your assessment of a large portion of your life ("Not that happy; I feel like I've been drifting for the last year or so"). When you say that people strive to be happy, then, you might be saying, as Freud does in the quote above, that people want to maximize pleasure and minimize pain. Or you might, like Gilbert and Le Guin, mean something more abstract.

Some well-known studies by the psychologist Daniel Kahneman and his colleagues tried to pull apart these different senses of happiness. Consider first what they call "experienced happiness." This is your experience of the psychological present; it's how you feel right now. If this was all that mattered, we could determine a life's value by just adding up the quality of each of these moments. To put some numbers down, if the psychological present lasts for about three seconds—a reasonable estimate from studies of memory and consciousness—the value of a seventy-year life is the sum of about half a billion moments. (We're just counting waking moments; let's save the question of the happiness or sadness of sleeping people for another day.)

There are practical issues that arise here. Suppose you just wanted to measure the sum of your experienced happiness for a year—about seven million moments. It would be seven million very boring moments if they were all spent answering the question "How are you doing?" So instead you can extrapolate from random samples over that period. The collection of these samples can be done through a smartphone app that randomly goes off; when it does, the volunteer answers questions about how he or she is feeling. Or, as was done by Kahneman and colleagues, you can ask people each morning about their previous day—questions such as "Did you experience the following feelings during a lot of the day yesterday? How about _____?" with emotions like "stress," "happiness," "enjoyment," "worry," and "sadness" filling in the blank. This measure is contaminated by bias and memory, but it roughly captures the notion of moment-to-moment happiness. And, again, you can figure out the value of a year, or of a life, by adding up your results from these individual day measurements.

This is experienced happiness. Now consider a different judgment one can make, what we can call satisfaction. This is a more contemplative assessment, looking at what you think about your life as a whole, not immediate moments. One method used to test this is the Cantril Self-Anchoring Scale, where there's a ladder and you mark down where you stand between 0 and 10, where 0 is "the worst possible life for you" and 10 is "the best possible life for you."

What is the relationship between experienced happiness and satisfaction? Daniel Kahneman and Angus Deaton did a survey of a thousand U.S. residents, accumulating more than 450,000 responses, measuring both their day-to-day experiences and their overall satisfaction with their lives. Now, our minds could have worked so that the two measures converge on the same answer—one's judgment of satisfaction could just be an averaging of experienced happiness. But this wasn't the case.

Consider the effects of money. When it comes to experienced happiness, more money makes you happier. This makes sense. Money can buy you positive experiences and can make your life better in all sorts of ways. More to the point, being poor makes everything worse—as the authors put it, "Low income exacerbates the emotional pain associated with such misfortunes as divorce, ill health, and being alone."

There are diminishing returns, though. If you are making $30,000 a year, another $5,000 is a big deal, but if you are making $300,000, not so much. This makes sense; it's true of good things in general. Being friendless is rough, so it's a lot better to have one friend than no friends, and better to have two rather than one . . . but you wouldn't expect the same sort of jump when you have twenty friends and get one more.

It turns out that for experienced happiness, money matters only up to an annual income of about $75,000. (This study was done in 2010, so we might adjust that to $89,000 for inflation.) Apparently, the day-to-day experiences of a well-off person and a very rich person aren't that different, perhaps because the sorts of things that lead to experienced happiness, such as social contact and rewarding work and good health, don't necessarily become abundant as you get richer.

What about the effects of money on satisfaction? Just as with experienced happiness, money is related to satisfaction, and again there are diminishing returns. But here's the difference: While there is a threshold after which experienced happiness levels out, there doesn't seem to be one with satisfaction. There is no point in their study where more money isn't associated with more satisfaction. When the question is "How is your life as a whole?," the more money, the better.

This point is worth emphasizing, since there seems to be an urban legend that money, at least past a certain point, doesn't make much of a difference in the quality of your life or even makes you miserable. This just isn't so. Take a poll from 2019, which looked at people split into four categories: lower income (less than $35,000 per year), middle income ($35,000–$99,999 per year), higher income ($100,000–$499,999 per year), and top 1 percent (more than $500,000 per year). Most studies underrepresent people at the higher income levels; this one took care to include 250 respondents. And here is the proportion of each group that were "very" or "completely" satisfied with their lives:

Lower:	44 percent
Middle:	66 percent
Higher:	82 percent
Top 1 percent:	90 percent

It doesn't stop there: Another study looked at the superrich and found that people with more than $10 million in wealth were more satisfied with their lives (though just by a bit) than those with a mere $1 to $2 million.

Taken as a whole, these findings suggest that when we think about our overall lives, we tend to compare ourselves with others—and when it comes to social comparison, the sky is the limit. Along the same lines, Kahneman and Deaton find that health matters a lot for the experience of the present moment (there is something about being healthy or sick that affects you in a day-to-day way, regardless of the health of others), while degree of education matters more for satisfaction (which is consistent with a social comparison account).

NOW, WHEN SOMEONE suggests that people just want to be happy, you can ask: Which kind of happiness do they aspire to? Do they try to live so that each moment gives the most pleasure? Or do they want to maximize their overall satisfaction?

In a recent podcast with Tyler Cowen, Kahneman argues for the importance of satisfaction:

Cowen: One result from [your research] is how much people enjoy spending time with their friends. If that's so much more enjoyable at the margin, why don't people do more of it?

Kahneman: Altogether, I don't think that people maximize happiness in that sense. And that's one of the reasons that I actually left the field of happiness, in that I was very interested in maximizing experience, but this doesn't seem to be what people want to do. They actually want to maximize their satisfaction with themselves and with their lives. And that leads in completely different directions than the maximization of happiness.

Many people think this is what *should* matter. In his analysis of the research I described above, the journalist Dylan Matthews writes, "I think it's fair to say that this metric—life satisfaction—is a better gauge for what people actually want for themselves than emotional well-being is. I don't want to be perpetually giddy and worry-free; I do want to have a life that I'm, on the whole, happy with."

I agree with the gist of this. One theme of this book is that we are not only hedonists—we are not only trying to maximize our immediate pleasure—and it's a good thing that we aren't.

But I'm not sure that life satisfaction is all that we are looking for. Keep in mind that the big finding from the research is that when we aspire to a life that we are, in Matthews's words, "on the whole, happy with," we focus a lot on social comparison. Notably, we try to make more money than everyone else.

This sort of one-upmanship seems hard to defend and might be poor advice for a life well lived. Is there something besides this sort of happiness that we should be striving for? What else is on the table?

LET'S PUT ASIDE *happiness* for a moment, with all its vagueness and multiple meanings, and also give up on *satisfaction*, which includes good things such as purpose and meaning but also not-so-good things, like impressing everyone at your high school reunion. Let's go back to the question of what people want and consider an answer that, whatever else one might say about it, is at least pretty clear.

It's *pleasure*. The Greek term for pleasure is *hēdonē*, which is why those who argue for the centrality of pleasure are called hedonists. The spirit of this view is nicely captured in *The Epic of Gilgamesh*: "Let your belly be full, enjoy yourself always by day and by night! Make merry each day, dance and play day and night! . . . For such is the destiny of men." And also by the Canadian rock band Trooper: "We're here for a good time / Not a long time / So have a good time / The sun can't shine every day."

Hedonists wouldn't deny that life is full of voluntary suffering—we stagger out of bed at 3 a.m. to feed the crying baby, take the 8:15 into the city, undergo painful medical procedures, and so on. As Trooper put it, the sun can't shine every day. But for the hedonists, these unpleasant acts are the costs that have to be paid to obtain greater benefits. We are fated to reenact the biblical punishment of Adam, condemned to survive only by the sweat of our brow. Challenging and difficult work is the ticket to status and money; boring exercise and unpleasant diets are what you have to go through for abs of steel and a vibrant old age. To use the slogan of the libertarians, there ain't no such thing as a free lunch. Suffering is the price we pay for greater pleasure.

Many psychologists are hedonists, whether they admit it or not. They believe that pleasure is our ultimate goal. I see this in some of the responses to my research on morality. I have argued elsewhere that much of morality is bred in the bone. Even babies and young children have some concern for the fates of others, some interest in fairness and justice. But this initial morality is limited—selfish and narrow in the way that one would expect from the product of natural selection—so this moral foundation requires the right sort of personal and social experience to blossom into a more mature morality in adulthood.

At least that's the argument I've been making. Now, some other scholars believe that I'm wrong and that babies are moral blank slates, indifferent to the suffering of others and unable to tell right from wrong. I find the objections I've heard unconvincing, but I'm comfortable with the standard give-and-take here—maybe the experiments I'm relying on can't be replicated, or can be better understood in some other way, or perhaps new data (or new ways of making sense of old data) will challenge my conclusions. This is how scientific debate works.

But the response that surprises me is when someone says that babies can't have moral motivations. That they can't have them because *nobody* has them—there is no such thing. We might think we care about right and wrong, this view holds, or want to punish evil and reward good, or seek out fairness and justice and kindness, but the truth is, there's nothing more here than selfish drives. "Scratch an altruist," wrote the biologist Michael Ghiselin, "and watch a hypocrite bleed."

I don't want to mock this view; many brilliant people hold it. There is a story about Thomas Hobbes walking through the streets of London with a friend and then stopping to give money to a beggar. His friend challenges him, saying that Hobbes has long argued for the selfish nature of man, and Hobbes responds by saying that his action *was* thoroughly selfish—giving to the beggar gave him pleasure, and it would have made him feel bad to pass him by.

And then there is this story of Abraham Lincoln, as it was reported in a newspaper at the time:

> Mr. Lincoln once remarked to a fellow passenger on an old-time mud-coach that all men were prompted by selfishness in doing good. His fellow passenger was antagonizing this position when they were passing over a corduroy bridge that spanned a slough. As they crossed this bridge they espied an old razor-backed sow on the bank making a terrible noise because her pigs had got into the slough and were in danger of drowning. As the old coach began to climb the hill, Mr. Lincoln called out, "Driver, can't you stop just a moment?" Then Mr. Lincoln jumped out, ran back, and lifted the little pigs out of the mud and water and placed them on the bank. When he returned, his companion remarked: "Now, Abe, where does selfishness come in on this little episode?" "Why, bless your soul, Ed, that was the very essence of selfishness. I should have had no peace of mind all day had I gone on and left that suffering old sow worrying over those pigs. I did it to get peace of mind, don't you see?"

Under this view, our moral actions—our *so-called* moral actions—are just attempts to avoid the pain of guilt or worry.

FOLLOWING MOST PHILOSOPHERS, I find psychological hedonism implausible. I agree that we often seek out pleasure for its own sake, that we often scratch where it itches. But this isn't our sole motivation.

There are all sorts of specific goals we might have. Right now, as I write this, I hope the Blue Jays do well this season (the prospects are not good). I hope my younger son has a fun and safe (but primarily safe) trip through Nepal; I hope my older son does well on a job interview he's about to have. I'd like to make progress on this book and have a draft of the first half in the next three months. I would like the current president not to be elected for a second term. I have a friend coming out with a new book and I hope it's successful—he deserves it. I hope a certain person I've read about goes to prison. All of these specific motivations derive from more basic ones, but none of them reduces to a simple desire for pleasure.

Self-deluded! responds the psychological hedonist. After all, wouldn't it be a positive experience for me if these things come to pass and a negative one if they don't? Well, yes: part of what it means to want something is that you are pleased when it happens. But this isn't an argument for hedonism, because it doesn't show that the pleasure is the goal itself, as opposed to a by-product. If you ask a friend what time it is and she turns to you and explains that you don't really want to know the time—you just want the pleasurable buzz that this knowledge will provide—you need to get better friends.

Let's zoom in a bit on an everyday example: love for one's children. There's nothing unusual in wanting your children to thrive. This is true even when there's no tangible payoff, when you're not expecting them to take care of you in your dotage, for instance. Parents of a mentally disabled girl might spend many difficult hours each day working to give her a life of joy and dignity and some degree of autonomy. They might be careful with their money, giving up on certain luxuries, setting it up so that after they die, their daughter will be well cared for (even though they will not be there to see this). If you were to ask them why they are making these sacrifices, they would likely tell you that they love their daughter and want her to have the best life possible. This is a good explanation for all of these behaviors. You don't need to be a hard-core evolutionary psychologist to appreciate that there is selective pressure for animals to evolve to help their offspring thrive—and, for sophisticated creatures like us, one way to generate this help is through love. (While this motivation has evolved for biological children, it extends more generally, and so all this could occur if the daughter were adopted.)

The psychological hedonist might jump in at this point and tell the parents, "You're not *really* motivated by your love of your child. You just want to get the warm glow of helping her or avoid the pain of guilt if you abandoned her." But why should this be taken seriously? It's certainly not the felt experience of the parents. And it makes the wrong predictions. This hedonist alternative suggests that if the parents would get more pleasure and less pain from abandoning their child—perhaps there is a drug they could take that would extinguish the love and blot out any future guilt—they would make that choice in a second.

Now, some might say yes to this—there are those who prefer the company of drugs like heroin to everything in the world. But I bet most parents in this situation would not.

Or consider the soldier who opts to die to save his comrades by jumping on a grenade. Some choices to die are readily explained in hedonistic terms—to escape from pain, say. But not this one. And, no, not everyone who performs such an act expects eternal blissful reward in heaven—there are plenty of atheists in foxholes.

Again, I don't deny that hedonic motivation plays some role in our everyday lives, and I'm comfortable agreeing with the cynics that sometimes we fool ourselves into thinking otherwise. Studies of voting patterns, for instance, suggest that political positions align suspiciously well with self-interest. Want to know what Jane thinks about government-supported childcare and taxes on the rich? Well, it turns out that you can learn a lot by checking whether she has children and how much money she makes.

But this is far from the whole story. The evidence also suggests that we are wired, through natural selection and then through exposure to culture, to want our communities to be better places, to see justice done. This means that we have psychological motivations that are distinct from, and sometimes at war with, hedonistic ones.

What do we say about people (and I've met a few) who insist that they themselves are hedonists? They say that to the extent they do something for others, or engage in difficult long-term projects, they do it for the warm feeling they get. I'm not thinking of someone who is enjoying some lazy time at the beach, eating a hot fudge sundae, or in some other way taking a break from more difficult pursuits. I'm not even thinking of people who have reached a point in life where they want things to be easy—grandkids and crossword puzzles and good books by the fire. Rather, I'm thinking about those who say that they have no concern other than pleasure and insist that this has always been the case.

Maybe they're just wrong about themselves. As a psychologist, I have no trouble accepting that people can have mistaken theories about what's going on in their own heads. Freud was surely right that you can think you are doing something for one reason but actually be doing it for another.

I see this sometimes in my seminars on moral psychology, where my students and I explore competing theories of altruism, fairness, loyalty, vengeance, taboos about sex and eating, and so on. Often enough, when we're sitting around the seminar table during the first meeting, someone will say that they don't believe there really is such a thing as right or wrong. Sometimes the student is thinking about morality in a very narrow sense, identifying it with fundamentalist religious views; maybe sometimes he or she is just messing with me. Still, I do push back, and one way to do so is to ask what the student thinks about some policies I'm considering implementing for the rest of the semester. I tell them that I plan to give Black students lower grades and not to admit trans students into the course and to insist that the women leave the seminar room when we are talking about complex issues.

The class knows what I'm doing here, of course, but there is still a hiss as I speak, and at this point the student will typically admit that, yes, these plans of mine really do seem wrong. Not just impractical or unconventional or non-utility-maximizing, but wrong. The point here is that many of those who think they are morally uninterested can be pretty quickly reminded that this isn't really how their own mind works. (Actually, there's probably no human more concerned about morality—with all the good and bad that this implies—than the American college undergraduate.)

Still, perhaps true hedonists do walk among us. For anything that falls on a continuum, there will be those on the extremes. After all, people differ in their sex drives, and there are some who have no sexual interest at all. In much of my work, I've been arguing for natural moral motivations, but it's sometimes said that there are pure psychopaths without moral feeling. (None have revealed themselves in my seminar, but of course they wouldn't—such individuals wouldn't have made it very far in the world if they were open about their psychopathic natures.) And so it's possible that there are those who are truly indifferent to concerns that don't reduce to pleasure. But most of us don't work that way.

I PROPOSE THAT there are multiple independent drives that normal humans possess. Some are hedonic. This includes sexual pleasure, the satisfaction of hunger and thirst, and even the right sorts of relatively low-level pain. Others are moral, including a desire to do good, to be fair, to seek justice. A third, related class of motivations has to do with meaning and purpose. (The proper term for this is "eudaemonic," but it's an awful word to write and to say and I'll try to use it as little as possible.) This includes the pursuit of goals such as going to war, climbing mountains, and being a parent.

These motivations are plainly compatible. You can have a life that includes both pleasure and meaning, and, though meaning does involve suffering, a meaningful life doesn't have to be a grim one—there are activities that are stressful and difficult at times but also are quite a bit of fun.

How do we rank these different motivations? The philosopher Robert Nozick gives the example of an experience machine. Being plugged into the machine will generate the illusion of living a life of intense pleasure, happiness, and satisfaction. Concerned that you'll feel that you're missing out on the real world? No worries—the machine can blot out your memory that you are in the machine. It's some combination of the Matrix and Woody Allen's Orgasmatron, only better.

Nozick says that he wouldn't plug himself into the machine, and many people, including me, wouldn't either. We want to live in the real world; to do things, not just have the experience of doing things. Indeed, for Nozick, "it is only because we first want to do the actions that we want the experience of doing them." More generally, "someone floating in a tank is an indeterminate blob"—and who wants to live their life as an indeterminate blob?

I'll admit, though, that not everyone has the same reaction. This post on Twitter made me laugh, but some people do think this way.

Philosopher Robert Nozick: "Now this experience machine can perfectly simulate a life in which you get everything you ever want—"
Me: "Sign me up."
RN: "No, see, it won't be real; you'll think it is, but—"
Me, already plugging in: "Bye nerd."

After all, there are those who choose to take drugs that blot out any chance of meaning and authenticity, hoping to achieve hedonic bliss. Surely *they* would sign on to the machine.

Also, some skeptics have worried that the intuitions of people like me might be corrupted by a status quo bias, by a tendency to keep doing what we're used to. Since we're not now in the machine, entering it would be a shocking change. But imagine flipping Nozick's scenario: Suppose you are living a good and satisfying life—perhaps this is your life right now—and suddenly *poof!*, you find yourself in a white room, and some pleasant lab tech tells you that you have just spent the past few years in the experience machine. All your satisfactions, victories, and relationships are neural hallucinations. This is your regular check-in, mandated by the government, where they ask you whether you want to stay in the experience machine or return to the real world—which is, of course, far less enjoyable. If you decide to stay in the machine, your memory of this check-in will be wiped, and you'll go back to thinking that your life is real.

I'm honestly not sure what I would do in such circumstances. But some people I've spoken to would still leave the machine, even under these circumstances. This suggests that existing in the real world not only matters but, for at least some of us, matters more than a life filled with pleasure.

I've had little to say so far about what, precisely, I mean by "meaning." More is coming! But I want to end this introductory chapter by giving further support for motivational pluralism, more reason to distinguish a life of pleasure from a life of meaning.

Let's start with the work of Roy Baumeister and his colleagues, who did a series of surveys on hundreds of individuals. In one of the surveys, they asked about happiness (with seven-point scales) by asking people how much they agreed with the following statements: "In general I consider myself happy"; "Taking all things together, I feel I am happy"; "Compared to most of my peers, I consider myself happy." And they also asked about meaningfulness: "In general I consider my life to be meaningful"; "Compared to most of my peers, my life is meaningful"; and "Taking all things together, I feel my life is meaningful." (Now, given the problems with "happiness" we talked about earlier, this is less than perfect—I wish they had asked about "pleasure" or something else specific.)

Then, in other surveys, they asked the same people about various facets of their lives. This helps inform us about what kind of life is associated with people who see themselves as having a happy life, a meaningful life, both, or neither.

It turns out that some features of one's life relate to both happiness and meaning. If you describe yourself as being bored, then you are less likely to have either a happy life or a meaningful life. Similarly, if you describe yourself as lacking social connection—as lonely—this is also bad for both happiness and meaningfulness. Indeed, one main finding by Baumeister and his colleagues is that there are correlations between happiness and meaning: people who were high in happiness tended to say that they were high in meaning, and vice versa. You can have both.

Still, some people are high in one and low in the other, and there are traits that are related to happiness but not to meaning, and vice versa. Here are four differences.

1. Health, feeling good, and making money are all related to happiness but have little or no relationship to meaning.

2. The more people report thinking about the past and the future, the more meaning they say they have in their lives—and the less happy they are.
3. Finding your life to be relatively easy is related to more happiness; finding your life to be difficult is related to less happiness and, though it is a small effect, more meaning. Do you consider your life a struggle? You're likely to be less happy but more likely to see your life as more meaningful. Are you under stress? More meaning and less happiness. What about worrying? Again, more meaning and less happiness. These findings mesh with a study we'll discuss in more detail later, in which those who reported the greatest amount of meaning in their jobs included social workers and members of the clergy—difficult jobs that don't make much money and that involve dealing with complicated and stressful situations.
4. The researchers asked, without any elaboration, this simple question: "Are you a giver or a taker?" The effects are small here, but there is a pattern: Givers have more meaning in their lives; takers have less. Takers have more happiness; givers have less.

To sum up, happy people tend to be healthy and financially well-off, and to have lives with a good deal of pleasure. Those who find their lives meaningful might have none of this; they set ambitious goals, and their lives have more anxiety and worry. In a later discussion, Kathleen Vohs, a coauthor on the original article, writes, "The results revealed that happiness is about feeling good, avoiding feeling bad, and having one's own wants and needs met. By comparison, meaning in life was predicted by behaviors and feelings reflecting concern for others and outcomes, as evidenced by arguing, worry, and stress."

CONSIDER NOW ANOTHER distinction between meaning and happiness. In 2007, Gallup polled more than 140,000 respondents in 132 countries. They had the standard question for life satisfaction—people were asked to indicate where their current life stands on a ladder scale ranging from 0 (worst possible life) to 10 (best possible life). But they also had one other relevant question: "Do you feel your life has an important purpose or meaning?"

The happiest countries were the usual suspects—Norway, Australia, Canada, and so on. They are wealthy, secure, peaceful, with good social support. This survey, like the others, found that life satisfaction is strongly correlated with GDP per capita.

But in contrast, the countries where people reported the most meaningful lives included Sierra Leone, Togo, Senegal, Ecuador, Laos, Chad, Angola, Cuba, Kuwait, and the UAE—many of which had little wealth, security, or peace. Indeed, GDP had a *negative* relationship with meaning. The poorer the country, the more likely people were to say that their lives had an important purpose or meaning.

How can we explain this? Participants in this survey were also asked "Is religion an important part of your daily life?" and it turned out that self-reported religiosity correlated with meaning. Since religion also correlates with poverty, you might therefore get an indirect relationship between poverty and meaning.

Or perhaps poverty itself has a more direct relationship to meaning. Discussing these findings, Adam Alter suggests that "perhaps because poverty strips people of happiness in the short term, it forces them to take the long view—to focus on the relationships they have with their children, their gods, and their friends, which become more meaningful over time." To put it another way, perhaps when life is comfortable, you have a better chance at escaping struggle. If struggle is related to meaning, as I will argue in later chapters, then this would explain why affluent countries, particularly those with a strong welfare state, have citizens whose lives are relatively lacking in higher purpose.

I've argued against simple hedonism early on, but these data should make us appreciate the value of pleasure and happiness. They should make us, if not pro-hedonist, at least anti-anti-hedonist. After all, where would you rather live, Norway or Chad? Would you rather settle in Canada or Sierra Leone? Perhaps there isn't necessarily a right answer to these questions, but if this is what the contrast between happiness and meaning looks like, I'd take happiness, and I bet most people, including many of the residents of Chad and Sierra Leone, would agree with me.

But I suggest that we can have it all. Remember that in the Baumeister study, happiness and meaning were correlated—having one ups the odds of having the other. Keep in mind as well that it's not as if people in the wealthy countries have lives bereft of meaning; for instance, two-thirds of the respondents in Japan and France—relatively happy and rich societies—told the pollsters that their lives had meaning. This is not small potatoes.

I'LL END WITH a Zen thought experiment. I was in a movie theater, waiting to watch *Avengers: Endgame*, and this commercial for a bank came on, but it hardly mentioned the bank at all.

Instead, a narrator read out a speech while pretty images flitted across the screen. When I went home, I googled phrases from the speech to see where it came from (it had the flavor of a literary quotation, not something thought up by an adman) and it turned out to be written by Alan Watts, the British philosopher and popular interpreter of Zen Buddhism.

Watts begins by asking you to imagine that you are able to dream about whatever you want, with perfect vividness. Given this power, you could, in a single night, have a dream that lasted seventy-five years. What would you do? Obviously, he says, you'd fulfill all your wishes, choose every sort of pleasure. It would be a hedonistic blowout.

And then suppose you can do it again the next night, and then the next, and then the next. Soon, Watts says, you would say to yourself:

> But now let's have a surprise, let's have a dream which isn't under control, where something is gonna happen to me that I don't know what it's gonna be.

And then you would continue to gamble, adding increasing risk, uncertainty, ignorance, deprivation. You would put obstacles in your way, obstacles that you might not be able to overcome, until finally, as Watts says,

> you would dream the dream of living the life that you are actually living today.

Is your life right now—with its difficulty and struggle, worry and loss—the best that life can be? Probably not. But Watts's fantasy is close enough to the truth to be profound.

2

BENIGN MASOCHISM

When was the last time you screamed? For me, it was a few months ago, in a hotel room in Mumbai. I was packing up early in the morning and tried to remove an adapter from the wall. It was a loaner from the hotel, an ugly thing with metal prongs in unpredictable places, and I must have touched it the wrong way, because I ended up flat on my back on the other side of the room, gasping and shaking. Later on, we'll talk about how the most normal of people can get perverse pleasure from mild jolts of electricity. But this wasn't mild at all, and, for just a second, I understood how electrical shock can be used as torture.

We scream when we are in pain. But, weirdly, we also scream for the opposite of pain—intense pleasure, joyous surprise, great excitement. Have you seen the videos of fangirls in the sixties in the presence of the Beatles? They positively *shriek*.

Crying is also triggered by opposites. You might cry on the worst day of your life and on the best. Weddings and funerals; the thrill of victory and the agony of defeat. I have a friend who sees himself as a bit of a tough guy, but I watched him tear up at a sappy commercial made for the Olympics about how our mothers help us when we fall down (see the notes for the link). And I sniffled along with him, though it's hard to put into words precisely what was squeezing the tears from us.

Crying is mysterious. One of my favorite books is *Pictures and Tears.* It's by the art critic James Elkins and it's all about paintings that make people cry. Sometimes the paintings depict awful events that might make you cry if you saw them in real life, such as the death of a child. Sometimes they have painful associations. Elkins heard from an English professor whose wife had had an affair. She had recently made a painting of their bed, empty and unmade, and one day when the professor was alone in the house, he looked at the painting and thought about what it meant, and he began to cry. But Elkins also heard from people who cried because the paintings were so beautiful that they could hardly bear it; they were moved by a positive emotional response to extraordinary human creations.

Once you look for paradoxical reactions, you see them everywhere. We laugh at what's funny, but we also laugh when anxious or embarrassed. We grin when happy, but sometimes we grin when angry. Smiling is associated with joy, but when researchers asked people to watch a sad movie scene—the part of *Steel Magnolias* where a woman is speaking at the funeral of her adult daughter—about half of the subjects smiled. Or consider the face associated with orgasm. It looks a lot like someone in agony, marked by scowling and grimacing. Actually, when shown photographs of people's faces during orgasm, subjects misidentified them as faces of pain about 25 percent of the time.

In general, extreme expressions are hard to interpret. The authors of a paper in *Science* give the example of two people, one of whom has just won an enormous lottery, while the other has just watched his three-year-old getting hit by a car, and suggest that, looking at their faces, you might not be able to tell them apart. In support of this, they find that people are unable to distinguish between winners and losers of high-stakes sports competitions when they see their faces in isolation (although, interestingly, once they saw their body postures and knew what the athletes were responding to, they would then "see" the emotionality of the faces—they were no longer ambiguous).

To take a different case, think about the reaction that people sometimes have toward babies. Filipino has a word for this—*gigil*—which refers to the agitated feeling many of us get toward the adorable and vulnerable. We want to pinch and squeeze. We often nibble on babies and say we are going to eat them. Just imagine, your friend shows you his one-year-old baby, and you lean over, grab the baby's toes, gnaw on them, and growl "I want to gobble you up!"—and nobody thinks you're crazy, not even the baby. In a survey done by Oriana Aragón and her colleagues, they found that most people agreed to statements like these:

> If I am holding an extremely cute baby, I have the urge to squeeze his or her little fat legs.
> If I look at an extremely cute baby, I want to pinch those cheeks.
> When I see something I think is so cute, I clench my hands into fists.

> I am the type of person that will tell a cute child "I could just eat you up!" through gritted teeth.

The theory that Aragón and her colleagues have about these strange reactions is that they arise when your feelings—toward the Beatles, an artwork, a baby—become overwhelming. You need to calm the system down, and so, to compensate, you generate expressions and actions that counteract your feelings, that go in the opposite direction. Think of it like cold water on a fire that might get out of control. The researchers who studied the orgasm face argue something similar, suggesting that the expression is an attempt to regulate a "too-intense sensorial input."

This sort of compensation—pursuing the negative to balance out the positive, and vice versa—might work on a broader level. It might explain how we choose to organize our day-to-day lives. Usually a day has pleasant activities and unpleasant ones, and to some extent we have control over when to experience the good and the bad—when to go out with friends and when to clean out the kitty litter. To explore how we prefer to organize the good and the bad, one study used a smartphone app to make real-time measurements of the moods and activities of 28,000 people over about a month. It turned out that people's choices showed what the authors called "a hedonic flexibility principle." When they were unhappy, they tended to do things that made them happy, like playing sports, and when they felt happy, they would do necessary things that brought them no joy, like housework. The positive and the negative sat in balance.

THE WORD "MASOCHISM" was coined by the psychiatrist Richard von Krafft-Ebing in the late nineteenth century, and it derives from Leopold von Sacher-Masoch. He was the author of *Venus in Furs*, a novel about a man who convinces a powerful woman to enslave him, and then she abandons him to enslave herself to another man. Krafft-Ebing used the term to refer to a sexual anomaly where the core fantasy is "of being completely and unconditionally subject to the will of a person of the opposite sex; of being treated by this person as a master, humiliated and abused."

While the word has kept its sexual connotation, it quickly grew to have a broader meaning. In an article called "The Economic Problem of Masochism," published in 1924, Freud wrote about sexual masochism, but he talked as well about moral masochism, in which someone seeks out suffering to relieve guilt—something we'll turn to soon enough. And, more recently, Paul Rozin has coined the term "benign masochism" to refer to certain types of voluntary pain and suffering, most of which don't have anything to do with sex.

There are many things that benign masochism is not. It doesn't include the choice of difficult life pursuits, such as deciding to have a child. It doesn't include activities that can damage one's body or cause severe pain—it's benign, after all. Getting yourself crucified, as some of the faithful do in the Philippines during Easter, is not benign masochism. The pleasure and pain of saunas is usually a good example of benign masochism, but you can go too far. In the 2010 World Sauna Competition, two finalists passed out after six minutes in 230°F (110°C) heat, suffering from burns and trauma. One of the men later died; the other was put into a medically induced coma and kept there for six weeks before he woke up with severe injuries. That's not benign masochism, either.

Rather, benign masochism refers to the choice to pursue

activities that are normally painful or unpleasant but not harmful. We sniff with curiosity at food we know to be rotten, touch a sore tooth gingerly with our tongue, press down on a sprained ankle. We watch movies that make us cower and cry. We eat spicy food and immerse ourselves in hot baths. Many psychologists do experiments that involve the infliction of harmless and painful electric shocks, and the odd thing is that you don't have to pay subjects gobs of money to participate; people, particularly young people, and especially young male people, like to get shocked. Not as intense as my Mumbai incident, but real pain, seemingly for its own sake.

WE GET SOME clues as to what's going on here when we return to the sorts of examples at the start of this chapter. For cases like screaming at the Beatles and sobbing at a wedding, it's clear that the positive and the negative are intertwined. This is an old observation. Plato describes Socrates rubbing his aching leg and saying, "How strange would appear to be this thing that men call pleasure! And how curiously it is related to what is thought to be its opposite, pain! . . . If you seek the one and obtain it, you are almost bound always to get the other as well." In modern times, many psychologists endorse an "opponent-process" theory of experience, whereby our minds seek balance, or homeostasis, so that positive reactions are met with negative feelings, and vice versa. The fear of skydiving is followed by feelings of relief and accomplishment, for instance.

Actually, all experience is understood and valued in the terms of contrast. The only good answer to the question "How are you feeling?" is "Compared with what?" When our experience is unchanging, it ceases to be an experience at all. We get used to the same old same old. The smell of cooking, the chill of a swimming pool, the hum of an air conditioner–they all disappear from consciousness.

Contrast is relevant even for something as basic as seeing the world. Do me a favor and, after you finish reading this sentence, just look at something for ten seconds–this book, your laptop, the cigar you're about to light, your loyal hound napping by your feet. Everything seems still, but this is an illusion: your eyes are darting around in tiny movements (microsaccades). Through the use of machines that track people's eye movements, you can set up an image that moves in unison with the eye movements, so it remains fixed on the retina. If you were to participate in such a study, you would have the experience, for the first time in your life, of looking at something without your eyes dancing all over it–real stillness. And what is this like? It's nothingness; the scene quite literally *disappears*. Experience requires change.

We respond to differences, not absolutes, and this means that something can become pleasurable not because of any stand-alone properties it has, but rather in contrast to the experience of the past. As one neuroscientist put it, "Because the brain grades on a curve, endlessly comparing the present with what came just before, the secret to happiness may be unhappiness . . . the transient chill that lets us feel warmth, the sensation of hunger that makes satiety so welcome, the period of near despair that catapults us into the astonishing experience of triumph."

If this all seems vague, consider the research of my colleague Robb Rutledge and his collaborators. In laboratory studies, they asked people to go through a series of financial choices that were either certain or risky, and every few trials they were asked, "How happy are you right now?" The main predictor of reported short-term happiness wasn't how much the subjects were making; it was how much they were making relative to their expectations. Momentary pleasure and pain are, at least in part, relative experiences.

Part of the story of benign masochism, then, is that we sometimes play with pain in order to maximize the contrast with future experience, so as to generate future pleasure. We engineer experiences in which the rush associated with the period immediately after pain's release is powerful enough to outweigh the negative of the original pain. And so the bite of a hot bath is worth it because of the blissful contentment that comes when the temperature is just right; the mouth burn of hot curry is pleasurable because of the shock of relief when you guzzle down some cool beer.

Sometimes the contrast that enhances pleasure comes from the comparison with an actual past experience, and sometimes it's from the contrast with expectations, as in the Rutledge studies. Consider also a series of studies published by Siri Leknes and colleagues, designed to explore what investigators called "pleasant pain." Their study involved putting subjects in a brain scanner and exposing them to a series of experiences of heat—mild, intense, or in between. Before any experience, they got a warning about what they could expect, but sometimes the warning would be incorrect. The big finding is that while normally the in-between amount of heat was judged to be painful, it would stop being painful if it was preceded by a warning to expect intense heat—then it was reported as *pleasurable*.

Now, one might worry that this reflects a confusion about language. Perhaps when describing the later experience as "pleasurable," the subjects were just expressing the view that it was better than they had expected. But this is where brain scanning comes in. Leknes and colleagues find that during this "pleasurable pain," there is increased activity in parts of the brain associated with reward and value (medial orbitofrontal and ventromedial prefrontal cortices), as well as considerably less activation in areas associated with suffering and anxiety (insular and dorsal anterior cingulate cortices). It seems to be an honest-to-God positive experience.

It turns out, then, that if you think something is really going to hurt and it hurts just mildly, the magic of contrast can cause this mild hurt to transform into pleasure. Now, I'll add the obvious here, which is that you're not going to get this effect if the pain gets too intense. If you think you're going to get a blowtorch applied to the back of your hand and the experimenter instead pokes you with the lit end of a cigar, you won't go *Whee!* But these experiments do suggest that it will hurt a bit less.

Other studies find that after experiencing pain in the laboratory, such as by having one of their hands immersed in freezing water, people report that subsequent experiences, like the taste of chocolate, are more pleasurable. Want some cake? Can I shock you right before you eat it—it'll taste better! These experiments are a bit weird, but the main idea is familiar: everyone knows that food never tastes so good as when you are hungry, lying on the sofa is blissful after a long run, and life itself is wonderful when you're leaving the dentist's office.

THIS IS THE contrast theory of why we choose to experience pain. It's like the old joke about the guy who was banging his head against the wall; when asked why, he said, "It feels so good when I stop."

I remember shoveling snow as a child in Quebec, and what I remember isn't the effort or strain (I was a child, after all), but the cold burning the parts of my face that weren't covered and the barbs of snow and ice getting into my boots and melting. But when I was done, my mom gave me some hot cocoa and then I climbed into a warm bath, and it was perfect bliss. A friend once told me about a hike she did in the British countryside—hours and hours spent with a companion after they got lost, and how they hadn't brought enough food and water, and it was getting dark, and they were starting to worry . . . and then they found a logging road and stumbled into a town and walked into a pub, and they sat outside in the dark with pints of beer and plates of fish and chips, chain-smoking and laughing. My friend's eyes lit up with such joy as she told the story—the experience so much benefiting from the suffering that preceded it.

Choosing to experience pain to enhance subsequent pleasure is a powerful trick, but it only works some of the time. The balance has to be right. There is sweet relief in putting your hand under cold water after you burn it, but it's not so good that it makes up for the shock of grabbing the handle of a hot saucepan in the first place. People don't really bang their heads against the wall because it feels good when they stop, and nobody goes to the dentist because the end of the experience is so liberating. My seven-year-old self enjoyed the aftermath of snow shoveling, but if I had insisted to my parents that they send me off into the snowdrifts because I was jonesing for the post-shoveling buzz, they would have brought me to a therapist instead.

The circumstances in which you get pleasure from pain are going to be rare. And this makes sense. As both Bentham and Darwin knew well, the hurt of pain is there to get us to stop doing certain things. We feel pain when we touch something hot because heat damages the body, which is ultimately bad for survival and reproduction. The whole point of this pain is that it gets us to snatch back our hands and swear and run to the sink to pour cold water over our singed fingers and cringe at the thought of ever doing it again . . . and hence take extra care to stay away from the saucepan next time. If we could override this—if our minds, through the power of contrast, could take any painful experience and flip it around to pleasure—then pain wouldn't do what pain is supposed to do. Our lives would be pleasurable from moment to moment but would be terrible in the long run. To put it another way, if hurting one's body were an overall positive experience, because the subsequent experience of the pain going away just felt so delightful, then we would spend our time purposefully injuring ourselves in countless ways, and we would never make it to puberty.

This is true for psychological pain and suffering as well. Think about experiences of humiliation, loneliness, regret, guilt, and so on. The function of all of these is to guide you away from certain activities by helping you anticipate the consequences (*If I say this I will feel awful, so I'd better not say it*) and to teach you a harsh lesson if you transgress (*That was terrible; I never want to do it again*). If someone's mind didn't work this way—if they, perversely, took pleasure in the pain of the death of those they loved or enjoyed the feeling of social exclusion or took delight in gnawing anxiety at some incoming catastrophe—then their incentives would be scrambled and their life would turn to dust.

And so, for benign masochism to work, certain conditions must be met. The pain has to be relatively brief. It has to quickly fade, providing the space for pleasurable contrast.

And the damage cannot be severe. Still, when you meet these conditions, contrast can be used to create pleasure. And so we've solved part of the puzzle of benign masochism: explaining what goes on when people immerse themselves in painfully hot baths and eat painfully spicy foods.

This trick of benign masochism is made easier by a quirk of how we interpret our own experiences, discovered by Daniel Kahneman and his colleagues as part of the happiness research discussed in the previous chapter. Think of someone who sits back and listens to a musical performance, spending an extended period in total bliss. But then something goes wrong with the recording, and for the final thirty seconds there is an awful screech. Our music lover might say that it tarred the whole performance, even though her actual subjective experience was almost entirely fine. To make up some numbers, suppose the experience lasted for an hour, and each moment except the ending was great, 10 out of 10. But the last thirty seconds was terrible, the worst, a zero. So actually, more than 99 percent of the performance was amazing, and the whole experience, averaged over time, is just the tiniest smidgen less than perfection. But that's not how it would feel. All she'd remember was the damn noise.

But if the performance started with the terrible noise and then went on for fifty-nine minutes and thirty seconds of bliss, it would be less of a problem. Similarly, a party that is mostly fine but ends with an embarrassing experience is remembered as substantially worse than if the embarrassing experience came at the beginning. Apparently, when we look back on an event, we don't focus on the sum of the experience but instead give extra weight to how the experience concluded.

In one experiment exploring this, researchers had people immerse their hands in freezing water for varying periods of time and then asked them which experience they wanted to repeat on a third trial—that is, which one caused them less pain. Here were the trials:

A. Sixty seconds of moderate pain.
B. Sixty seconds of moderate pain, then for thirty seconds the temperature is raised a bit—still painful, but less so.

Which event does it make sense to choose to have again? A, obviously, because A *has less pain*. And yet subjects prefer B, presumably because it ends in a not-so-bad way. To see how weird it is, imagine going to the dentist. (Kahneman's own real-world research tested people during colonoscopies, back when the procedures were considerably more painful.) You're lying back in the chair and having an excruciatingly painful procedure for a half hour, mouth open wide, fingers clenched on the armrests, sweating, and the dentist says, "Okay, we're done," and you say: "Could you do me a favor? I don't want to think back on this as a horrible experience, so would you mind giving me about five more minutes of mild pain?"

Isn't this weird? It illustrates the tension between our memory of an experience and the experience itself, how they might pull us in very different directions.

Most relevant for the purposes here, one lucky accident of this feature of memory is that pain-then-pleasure is recalled as better than pleasure-then-pain. Because of this, even if the amount of pain, taken in isolation, is the same as the amount of pleasure, if the pain comes first, the distortions of memory decrease the pain and increase the pleasure, improving the whole experience. My memory of the snow shoveling wouldn't be anywhere near as positive if the hot bath came first.

THERE'S AN OBJECTION that might have occurred to you while reading this. I've been saying that the pain during experiences of benign masochism is just regular old pain, as unpleasant as pain always is. It's experienced for the sake of subsequent pleasure, just as I might take on an unpleasant job to earn money to buy something that makes me happy later on. As the behavioral economist George Ainslie puts it, "The negative can be an investment in refreshing the positive."

But maybe that doesn't fit your experience. Perhaps you actually enjoy the painful part of pain-then-pleasure. You might like the burn of curry, the shock of the ice water plunge, the sadness of Adele's "Someone Like You." For you, the negativity is not a cost to be paid; it's valuable in itself.

One explanation for this involves anticipation. Perhaps, while you're feeling the objectively uncomfortable heat of the sauna, you're also anticipating, with joy, the thrill of diving into the cool Finnish lake. One of the positive features of pain-then-pleasure (which isn't present in pleasure-then-pain) is that you can enjoy thinking about the future payoff of pleasure while in the midst of the pain.

Or, to take an example we'll spend more time on later, think about the typical structure of revenge films. They begin with the hero at peace (after the death of his wife, John Wick bonds with a beagle puppy named Daisy to help him cope with grief), then there is some evil act that shatters everything (after a chance run-in with Russian mobsters, they break into his house, knock him unconscious, and kill his dog), and then we work toward the gratifying payback (the legendary hitman comes back from retirement and, "blind with revenge, John will immediately unleash a carefully orchestrated maelstrom of destruction"). The murder of Daisy is upsetting to watch, but since you know what sort of movie this is, the sadness is balanced by the thrill that soon you're going to see those scummy Russian mobsters *get what they deserve*.

I don't think anticipation is the whole story, though. There are other reasons to choose to suffer besides an investment in future pleasure. These include, among others, the satisfaction of morality, of knowing that your suffering is for a good cause, and the pleasure of mastery—the satisfaction that comes from control and accomplishment and autonomy in the face of difficulty. We'll delve into all of this in later chapters.

WE'VE TALKED SO far about contrast, but there's more. Another force that can make pain valuable is its power to focus the mind. Whatever the negatives of physical pain—or of emotions such as horror and disgust—they sure are attention grabbers. As Samuel Johnson put it, "When a man knows he is to be hanged in a fortnight, it concentrates his attention wonderfully."

In a recent article, Winfried Menninghaus and his colleagues make the case that the ugliness of some art—the grotesqueries of Francis Bacon or Lucian Freud—arises in part as an attempt to capture our attention through an unpleasant jolt, to make these artworks stand out from the rest. Violence in movies, sometimes shocking violence, is another case of this.

Negative experiences can focus one's mind in a particularly rewarding way. Psychologists who study benign masochism like to quote a dominatrix who said, "A whip is a great way to get someone to be here now. They can't look away from it, and they can't think of anything else." Rumi, the thirteenth-century Sufi mystic, agreed, asking, "Where is indifference when pain intervenes?" (Elsewhere he wrote: "Seek pain! Seek pain, pain, pain!") This has its appeal: pain can relieve anxiety by distracting you from your consciousness. It gets you out of your head. This is one regard in which sharp and sudden pain resembles what might seem to be its opposite—orgasm.

It's sometimes even said that pain can temporarily obliterate the conscious self. This sounds scary—or maybe a bit silly—so to get a concrete sense of what this can mean, let's turn to a particular form of chosen pain, BDSM (short for "bondage and discipline, domination and submission, and sadism and masochism").

Roy Baumeister makes the case that when one is engaging in sexual masochism, "awareness of the self as a symbolic, schematic, choosing entity is removed and replaced with a low-level awareness of the self as a physical body." As he sees it, sexual masochism falls into the same category as extreme exercise and getting drunk.

Why would you ever want to escape from your self? Well, as Baumeister points out, self-awareness carries a burden. In everyday life, you need to make decisions that you're responsible for, often disappointing others. You need to put a good face forward to the world; you have to manage your desires and deal with disappointment and guilt and shame. You're stuck with your memories, your worries about the future, and your anxieties about the immediate present. You are left with that same internal monologue, maybe a bit whiny, that you have had for a very long time. It's not hard to see how we could become sick of ourselves—not just sick of the bodies we occupy (though this might happen, too), but sick of our consciousness. When it comes to this sort of misery, there is a great truth to the classic breakup line "It's not you; it's me."

And so one of the joys of immersing yourself in certain activities, such as hard exercise or a difficult puzzle or being whipped, is that you lose the feeling of being conscious of yourself. You just *are*. It's often said that getting to this state is one of the goals of meditative practice, but for some of us novices meditation has the opposite effect. Being trapped in one's own head with no distractions can be a miserable experience—one's *me-ness* can be annoyingly salient. In contrast, the first time I "rolled" (sparred) with someone in Brazilian jiu-jitsu, I realized afterward that during that period, I thought of nothing else, my self was gone, and there was a sort of bliss to that. Indeed, I was once violently mugged on the streets of New Haven, and while it wasn't an experience I recommend, I did realize afterward that during the mugging, I had been in the moment. My mind didn't wander at all. Now when I want to distract myself from my own consciousness, I don't meditate; instead I listen to podcasts—there is an involuntary and automatic draw to the voices of others that can take you out of your head and finally shut down that internal *me, me, me.*

And this brings us back to pain. Pain can be better than meditation, because while meditation requires the constant choice to engage with the monkey mind, to gently push away those distracting thoughts, pain does the trick for you. If you think a podcast can be distracting, just try a whip. I wouldn't deny for a moment the terribleness of so much of pain, and I'm all in favor of interventions to make it go away—I'll have my dental surgery with anesthesia, thank you very much—but its distracting force cannot be denied, and for some people, under some circumstances, this positive can outweigh the negatives.

I am aware that what we're discussing—the infliction of physical pain, humiliation, and enslavement—can be, in different circumstances, the worst acts that people can do to one another. Elaine Scarry, in *The Body in Pain*, makes the same point as Baumeister in her discussion of torture, describing in detail how it serves to obliterate the self, destroying awareness and meaning. But Scarry discusses this in a context that is horrific.

What distinguishes torture from masochism? In torture, the intensity of the assault on the self can be more severe, limitlessly so. But this isn't the key difference. What really matters is choice. There are no safe words in torture. To *voluntarily* obliterate one's self, temporarily and under situations of control, is one thing, and it can be blissful. To have someone else do it to you against your will is an act of astonishing cruelty.

The importance of control is so often missed. When then-secretary of defense Donald Rumsfeld was asked to approve the practice of forcing prisoners in Guantánamo Bay to stand for many hours a day, he responded by noting that he also stood for most of the day (he had a standing desk). So how bad could it be? After reports of torture by American soldiers during the same period, some journalists, including Christopher Hitchens, decided to get waterboarded so that they could report what it felt like. These adventures were in part motivated by genuine curiosity and moral concern, and I don't think they were useless. But any such experiment, by its very nature, cannot simulate the real thing. The physical sensation of drowning is terrible enough, but surely part of what's so terrifying about being waterboarded is that the people who are doing it to you won't stop when you ask them to. Control and consent are morally essential and experientially critical.

I'M PRESENTING BDSM here as a normal expression of normal desires. Is this right?

Freud viewed certain nonsexual forms of masochism as part of normal life, but he argued that sexual masochism reflects mental illness (*perversion*, as he put it) and suggested that the participants in these interactions are actually sadists turning on themselves. Subsequent scholars in the psychoanalytic tradition linked masochism with criminality, epilepsy, pederasty, necrophilia (!), and vampirism (!!), with one describing the company of individuals who engage in such practices as the "kingdom of hell."

It would help if we knew more about the practitioners, but good studies are hard to find, and the numbers are all over the place. In one telephone survey, the experimenters asked if respondents had been involved in BDSM in the past twelve months. Only 1.3 percent of women and 2.2 percent of men said yes. That's a very small proportion. But if one asks about sexual fantasies, as was done in a study in 2015, 65 percent of women and 53 percent of men reported fantasizing about being dominated sexually—and 47 percent of women and 60 percent of men reported fantasizing about dominating someone else sexually. It's likely that there are a range of preferences here, from none at all to spoken fantasies during sex to mild faux restraints to (consensual) choking, spanking, and hair pulling, moving all the way to dungeons and whips and fire, and perhaps more.

In any case, if BDSM is a form of mental illness, it should be associated with other psychological problems, and it doesn't seem to be. Participants may actually have lower-than-average levels of depression; they tend to be extroverted, conscientious, happier, and, not surprisingly, open to new experiences. The only negative traits associated with BDSM are that practitioners may have higher levels of narcissism and lower levels of agreeableness than non-practitioners.

The mildest form of BDSM is when you don't actually experience it; you just imagine it. Consider *Fifty Shades of Grey*, which tells the story of the relationship between a young, beautiful, and fairly innocent student named Anastasia Steele and a young and beautiful man, Christian Grey, who is rich and not at all innocent. Grey introduces Ana to BDSM and, to use the sort of descriptive language you'll find in the book itself, "together they explore Christian's dark past, revealing deep secrets and steamy sex." People really do enjoy this story. *Fifty Shades* was the best-selling book of the 2010s. The second-best-selling book? The sequel, *Fifty Shades Darker.* The third-best-selling-book of the 2010s? The last in the series, *Fifty Shades Freed*. The accompanying movie versions also did well.

In this series, the reader is meant to take the place of Anastasia. The popularity of the books and the movies would have made Freud happy, as he insisted that masochism was part and parcel of the female condition—it's what women really want. Other commentators, such as Katie Roiphe, see the appeal of such fantasies as a reaction to the increased opportunities that women now have in the modern world: "But why, for women especially, would free will be a burden? Why is it appealing to think of what happens in the passive tense? Why is it so interesting to surrender, or to play at surrendering?

It may be that power is not always that comfortable, even for those of us who grew up in it; it may be that equality is something we want only sometimes and in some places and in some arenas; it may be that power and all of its imperatives can be boring."

This is an intriguing suggestion, consistent with the claims by Baumeister about escaping from the self. But, as often happens with this sort of social commentary, it's data-free. Roiphe links the appeal of *Fifty Shades* to a certain time (she wrote this passage in 2012), but she provides no evidence that *Fifty Shades* would have been less popular if it had been published fifty or a hundred years earlier, when women had much less freedom.

Also, Roiphe frames her article in terms of female desire, and yet these fantasies exist in men as well. I wonder how well a book would do that was just like *Fifty Shades* but with the genders reversed—an innocent young and beautiful man falls under the thrall of an older successful businesswoman with a taste for sexual domination. Maybe the dominatrix is in upper management at Google! Would this book be as successful as *Fifty Shades*? Is there a male appetite that is as yet untapped? If this idea motivates you to write this book and make millions, please remember that you heard it here first.

I've been defending BDSM as the reflection of normal appetites. This doesn't mean that everyone should do it, or that you're unhealthy if you don't do it—a desire for spicy foods is also normal, but it's cool if you don't indulge. But not all chosen pain is healthy. There are harmful behaviors that need to be distinguished from BDSM, though they do share some of its properties.

One such behavior is self-harm, or, more technically, NSSI—non-suicidal self-injury. This involves purposeful damage to the body (as opposed to the infliction of mere temporary pain) but not with the intent to kill oneself. It tends to start in the teen years: somewhere between 13 percent and 45 percent of adolescents report self-injury at some point in their lives. (As always, estimates vary depending on the precise question and who is being asked.)

Self-injury often involves cutting yourself with an instrument like a knife or razor, typically on the arms, legs, and stomach. There is also scratching, scraping, burning, and the insertion of objects like safety pins into and under the skin. These are the typical, relatively common manifestations, but there are also extremes. In Armando Favazza's *Bodies Under Siege*, he talks about finger amputation, testicle crushing, and eye gouging; the middle section of the book has chapters titled "The Head and Its Parts," "The Limbs," "The Skin," and "The Genitals." (And, yes, the book does contain pictures.) There is the biblical story of a man possessed by a demon who would "cry out and cut himself with stones"; he was subsequently cured by Jesus through exorcism.

Just like BDSM, NSSI involves chosen pain. But for NSSI, it is solitary, it is not sexual, and it is a response to serious problems in the person's life. It is not a pleasurable choice. People self-injure when experiencing feelings such as self-hatred, often in response to some traumatic event. In one study of inpatients, the decision to self-harm turns out to be very quick, coming in a matter of seconds. Drugs and alcohol are typically not used at the time of the event; the minds of self-injurers are clear.

I've described this in terms of pain because that's what usually comes with being cut, burnt, or otherwise injured. But, interestingly, people who self-injure report feeling little or no pain while in the act. And there are several laboratory studies that find that those who self-injure have lower-than-average sensitivity to pain—it takes them longer to say that an experience is painful, and they have greater tolerance for painful experiences. This might be because people who cut themselves, perhaps by nature, have unusually high pain tolerance, and that makes self-injury more appealing to them. Or perhaps this relative immunity to pain might develop as a reaction to repeated self-injury; cause and effect are hard to pull apart here.

Why do they do it? One answer is that it makes them feel better. One woman put it like this: "As the blood flows down the sink, so does the anger and the anguish." It's a distraction, an escape from the self, in much the same way as BDSM is said to be. In fact, one form of treatment involves replacing the damaging behavior with less harmful equivalents, like holding ice or snapping one's wrists with a rubber band.

But there's something else as well. NSSI is often motivated by a desire for self-punishment. People who do this sometimes want to discipline themselves for perceived wrongs. Self-injurers have carved words into their skin like "loser" and "disgrace." And when asked why they injure, they often say that they want to punish themselves. We should take them at their word.

Self-punishment is not a commonly accepted practice in modern communities. If I badly hurt someone and couldn't make amends, I wouldn't go up to a close friend and ask him or her to slap me in the face. But in other, more overtly moralistic times, things were different. Historian Keith Hopkins tells of second-century physician Galen's account of how a friend of his, overcome with rage, nearly beat one of his servants to death with a sword for a minor mistake. While punishment of slaves was accepted in Roman law, excessive punishment was illegal, and murder was still murder. Galen's skill as a doctor saved the slave's life and the incident passed, but later, "Galen's friend was overcome with remorse. One day soon afterward, he took Galen to a house, stripped off his clothes, handed Galen a whip, went down on his knees and asked to be flogged. The more Galen tried to laugh it off, the more persistent his friend became." Galen ultimately agreed to whip his friend, but only if the friend would first listen to a lecture by the physician on the virtues of self-control and the value of controlling his slaves through means other than violence.

Many years ago, I was having tea with a young merchant in Egypt. I had just paid far too much for some tchotchke, and we got to talking and I asked him, since our purchase was done, what was the most money he had ever gotten out of a tourist, not including me. He told me about an older European woman who came to his shop, and how she simply had no idea how much things were supposed to cost. He convinced her to buy a rug for more than ten times what any sensible person would have agreed to pay. And then he felt so awful afterward that, for the next day, he lived off water and salt.

You can explore self-punishment in the lab. One study asked people to write about a time when they had behaved unethically by excluding another person. Then they were told that they would participate in another activity, which involved the painful act of immersing their own hand in an ice bucket for as long as they could. (Psychologists love this—it's painful, but harmless.) The exclusion group held their hand in for longer than a control group that wasn't made to feel guilty, and they reported feeling better as a result. In a similar study, people were asked to write about a past event that made them feel "most guilty," and then were asked to manipulate a shock machine to either increase or decrease a set amount of shock they were receiving. Again the guilty group gave themselves more shock than a control group, and the stronger the shock they gave, the more their guilt went away.

A FURTHER REASON to seek out pain is signaling. Why did the man in Egypt tell me his water-and-salt story? Was he just lying to impress me and make another sale? (Perhaps by taking him seriously, I was a sucker twice over.) Or perhaps he was honest and he had actually done what he said, but the purpose of his sacrifice—consciously or unconsciously—was to show others what a good person he is.

Borrowing from animal research and evolutionary psychology, some scholars suggest that many of our reactions, tastes, and behaviors are best understood as ways to advertise positive aspects of ourselves to other people. The pleasure of suffering can be a social pleasure.

What aspect of ourselves are we advertising? It could be all sorts of things. One is toughness. The ability to withstand pain and turmoil—and the gumption to actually inflict it on oneself—can advertise one's physical and psychological fortitude. Or at least this is the idea: During a difficult point in physical therapy for an injured leg, I told my therapist, "I bet a lot of people have to stop at this point because they can't take the pain." She laughed and said, "No, but a lot of men say this to me." Such chosen suffering isn't necessarily a grim affair. After I talked about suffering on a podcast, Fernando Sánchez Hernández, a graduate student at the University of Illinois, wrote to tell me that part of the nightlife in Mexico City is a game called "toques," where people clutch metal tubes and see how long they can cope with electric shock. This is part of having a good time, the stuff of friendly competition with friends and family.

Self-injury is another area where signaling might play a significant role. Maybe, in addition to its other functions, self-injury is also a cry for help, a signal of distress, as in the famous description of self-injury as a "bright red scream."

One version of this theory was developed by Ed Hagen and his colleagues, who explore self-injury as a costly signal. Sometimes it's valuable for a creature to signal some truth about itself to others—perhaps about how strong it is or how smart or how dangerous. The problem with many signals, though, is that anybody could convey them, so there's no way for the audience to distinguish the truth tellers from the fakers. The solution here is costly signals—those that only truth tellers have enough resources or enough motivation to transmit.

Suppose, for whatever reason, you want people to know that you are rich. You could just say, "Hi, I'm a millionaire," but there are two problems with this. First, you might want people to know that you are rich but not want them to know that you want them to know that you are rich. Second, people who aren't rich can also say that they are rich, so people might take your pronouncements skeptically. What you need is some way of expressing your wealth that's seemingly accidental (and thus shows that you are rich without also showing that you want people to know it) but also can be done only if you actually are rich.

The standard solution is to display an expensive object on your body, such as a Rolex watch. This solves both problems.

First, you have plausible deniability: it's not obvious that you are doing this to announce your wealth (maybe you just like Rolexes). And second, and more to the point, this is the sort of thing that a rich person can do but a poor person can't. Indeed, one might argue that for some luxury products, their high price is their very point; if the price of a Rolex dropped too much, the company might go out of business.

Why are children in elite high schools taught Latin, Greek, or even Sanskrit? Some people will insist on the importance of this knowledge, but the signaling theorist will say that it's the *unimportance* that matters here. Having your children spend valuable time on material of no tangible utility announces to the world that you are free of material need; poor children have to learn useful material, so they can't take the hit. Why is there hazing for so many fraternities, elite military units, and street gangs? Because undergoing the hazing is a costly signal of interest in the group. If entering the group required a thumbs-up and a five-dollar entry fee, anyone could do it; it wouldn't filter the dedicated from the slackers. But choosing to go through something humiliating or painful or disfiguring is an excellent costly signal, because only the truly devoted would want to do it. So, too, for religious rituals: how much commitment does it show for a man to circumcise himself or his baby son—particularly before anesthesia was invented? *A lot.*

Now imagine you want to convince others that you need help, support, and love. If those around you feel love for you, you can simply ask. Or you can cry, which is a universal distress signal. Or, in some relationships, just mope and look sad, and pretty soon your parent or partner will come up to you and ask if they can make you feel better.

But what if you're less lucky? What if the people around

you are indifferent and stingy in their affections, perhaps suspicious that you are trying to exploit them? Or perhaps you are in the sort of relationship where there are opposing interests. (Hagen and his colleagues give the example of a daughter wanting her mother to protect her from an abusive stepfather, but the mother is loyal to her husband.) Now just asking or hinting won't do the trick. Now you would benefit from a costly signal.

Perhaps self-injury is such a signal. It's the sort of thing that someone would do only if they were serious. It's similar in attention-getting power to threats to commit suicide. But threats can be faked and lose force over time, while self-injury carries an inherent cost and so is a more credible signal of need.

I am intrigued by this idea, but also skeptical. For one thing, this theory predicts self-harm only in certain situations. If adolescents have good relationships with the adults around them, they shouldn't have to self-harm, and if they have terrible relationships and couldn't expect help under any circumstances, they also shouldn't self-harm—you don't signal if nobody is paying attention. This suggests that self-injury should be most common in cases that fall in between—the parents are not so loving that they will help if you just ask, but they are loving enough to help you if they can be convinced that you are in real trouble. There is, as far as I know, no research that tests this idea.

Also, cutting is typically done on hidden parts of the body. If you want to send a message, why not cut the face, the neck, or the hands? Hagen and his colleagues are aware of this challenge and have a couple of responses. One is that, for them, this sort of costly signal is an evolved mechanism, not a conscious strategy, so perhaps people have a compulsion to self-injure but then consciously choose to hide it. (Crying evolved as a distress signal, after all, and yet we often hide our tears.) Another possibility is that the point of such a signal is that it isn't seen as a signal—the self-injurer doesn't want to be seen as scheming and manipulative. So their injuries have to be discovered, not shown. We see this in more typical cries for help in a relationship. When I want attention from my partner, I might not want to be seen as wanting attention, and so I sulk, waiting to be asked, "Honey, what's wrong?" (The proper response: "Oh . . . nothing.")

In the end, costly signaling is unlikely to be a complete explanation for self-harm, but it might well sit alongside the other forces described above: mood regulation (escape from self) and self-punishment.

We can learn a bit about the pleasures of pain by thinking about what *can't* be transformed into pleasure. Pretty obviously, people tend to avoid severe pain and permanent bodily damage. Psychologically healthy people don't shoot themselves with nail guns or set parts of their bodies on fire. Few of us are true masochists, cooking up recipes from "The Masochist's Cookbook" (from the satirical magazine *McSweeney's*):

CINNAMON SPICED PECANS
WITH ORANGE RUM GLAZE

2 1/2 cups raw pecans
1 cup rum
2 tsp. light-brown sugar

1/4 tsp. salt
1/2 tsp. ground cinnamon
zest of 1 orange
Toast pecans at 350 degrees for 5 min. In a large saucepan, add rum, sugar, salt, cinnamon, and zest. Bring to a rolling boil. At this point, you may want to call 911. Remove pants. Bite down on an oven mitt and pour scalding glaze mixture over genitals. Serves: 4

There are also psychological experiences that people avoid. We often revel in terrible experiences in the safe space of fantasy—this is coming up in the next chapter—but nobody wants to go through a period where they actually believe that their child has been killed, that their friends hate them, that their deepest secrets have been revealed.

There are also more subtle limitations. Few people seek out nausea. In a *New Yorker* article on the topic called "A Queasy Feeling," Atul Gawande notes that Cicero described how he would rather die than suffer the tortures of seasickness (and after a long hot afternoon on a boat off Boston Harbor, I'll be damned if I ever go whale watching again). Some mothers have worse memories of the nausea of pregnancy than childbirth. If you sprain your ankle running, once you heal, you're back on the trail. But nausea is different—if a certain food makes you sick, it's hard to return to that food again.

Gawande notes the evolutionary logic here. The function of nausea is its role in expelling toxins you may have ingested; the awfulness of the experience, and the persistence of the awful memory, keep you from ingesting them again. This fits with the standard evolutionary theory of morning sickness and why nausea is so prevalent in the first trimester—this is precisely when the fetus is most vulnerable to natural toxins, so there is high vigilance.

The awfulness of nausea isn't enough, though, to explain its exclusion from the benign masochism menu. We like other awful experiences, after all. A better explanation is that nausea isn't amenable to the sorts of balancing tricks we talked about earlier, because it lingers. Pain that ends suddenly can set the stage for subsequent pleasure—I mentioned earlier the Finnish practice, which I would recommend to everyone, of stepping out of a hot sauna and diving into a cool lake. But an unpleasant experience that gradually fades over time doesn't set you up for a contrast; you can't enjoy the subsequent pleasure. Perhaps this is the problem with nausea. Suppose you had a nausea machine that caused extreme vertigo, but then, when you threw the switch to off, you felt totally fine. I bet the college students would line up to give it a try.

Boredom is a final example of something unpleasant that's hard to like. There are haunted houses where monsters jump at you and go *boo*, but there are no boredom houses where they take away your phone and there's nothing to read and you have to just sit for a couple of hours. In the study that I talked about earlier by Baumeister and colleagues, it turns out that being bored lowers both your happiness and your experience of meaning. There really is little to be said for it.

It is surprisingly hard to come up with a theory of what bores us. One early account was that we are bored when we are understimulated and underaroused—but then some later studies found that we can be bored when we are highly aroused. (Being stressed and being bored are compatible.)

Another theory is that we are bored when we feel we've lost autonomy. A few years back, passengers at a Houston airport would complain about how long they had to sit around and wait for their luggage. Rather than, say, get the luggage out quicker, the executives who ran the airport decided to move the baggage claim area farther away from the gates. Now, instead of being stuck with no way to predict or control when they got their bags, passengers would spend their time walking, pursuing a goal, and, the story goes, they stopped complaining because they weren't bored. Wouldn't you rather have a trip that involved an extra hour on the open highway than one where you spend an hour stuck in traffic?

But this autonomy theory can't be entirely right. If the walk to the luggage was long enough, people *would* get bored, even though they had autonomy and were working toward a goal. You can be autonomous and bored at the same time, and much real-world boredom is just like that: you can do whatever you want; you just can't find anything that interests you.

Later on, I'll discuss the importance of meaning in our lives, and it would nicely fit with one of the main themes of this book if it turned out that we are bored only with activities that lack meaning. But this plainly isn't true. Some meaningless activities—simple video games like *Angry Birds*, goofy sitcoms, trashy novels—aren't boring at all; they're cures for boredom.

Let's try another approach, then. We can get some sense of what boredom is by asking what it's for. Andreas Elpidorou sees boredom as being akin to pain. As we discussed earlier, insensitivity to pain is a curse. People with this condition will chop, burn, smash, and crush parts of their bodies and never notice that they did so; they need deliberate conscious thought to keep themselves intact. Pain is more of a motivator than rational deliberation. If you leaned against a hot stove and felt no pain, you might reason, *I should move away from the stove, because this damage to my body could be long-term and serious.* With pain, you scream and jolt back as fast as you can, tears in your eyes. You couldn't stay there for a million dollars!

The story of boredom, Elpidorou argues, is similar. Boredom is a cue that needs aren't being met. It's a signal that your environment lacks interest, variety, and newness. Just as the pain of a burn tells us where the damage is and motivates us to respond appropriately, boredom motivates us to seek out intellectual stimulation and social contact, to learn and engage and act. To be without boredom would be a curse.

Often, we don't allow ourselves to be bored, and we take advantage of technologies that engage our interest without providing anything of value. I do this myself—I lost about a year in graduate school to *Tetris*—but I see this as ultimately the wrong strategy, the equivalent of numbing the limb rather than pulling it from the heat. The best recent scientific article on the topic of boredom ends by making a similar point:

> [Boredom] is a canary in the coal mine of everyday existence, signaling whether we *want* and are *able* to cognitively engage with our current activity—and impelling us to action when we do not or cannot. How we respond to boredom matters: blindly stifling every flicker of boredom with enjoyable but empty distractions precludes deeper engagement with the messages boredom sends us about meaning, values, and goals. Empty maladaptive responses, such as self-inflicted

electric shocks in the lab, compulsive social media use, or full-scale gambling and drug use, may work to temporarily alleviate boredom, but at what cost?

WAIT—"SELF-INFLICTED ELECTRIC SHOCKS in the lab"? This refers to a clever series of studies reported in *Science* a few years ago. The investigators started by asking undergraduate subjects to surrender all of their belongings, including their phones and any writing material, and spend time sitting in an empty room, with only one rule: they had to stay awake. They were stuck there for between six and fifteen minutes, depending on the precise study.

The big finding was that they really didn't like it. The same held when older people were tested and when people were asked, using an online method, to do this at home.

How much did they dislike it? In one study, subjects were given experience with a painful electric shock and then asked how much they would pay to avoid getting the shock again. Then they were kept by themselves for fifteen minutes, with the shock machine in the room. Even though they said they would pay to avoid the shock, many of the subjects chose to indulge in the pain. (There was a big sex difference here. Two-thirds of the men shocked themselves—usually just once, though one man shocked himself 190 times—while only one-quarter of the women indulged.) This is yet another source of benign masochism—pain to escape boredom.

One reason why doing nothing is so unpleasant is that our thoughts, unfettered by distraction, take us to uncomfortable places. Boredom is the opposite of BDSM: instead of escaping from the self, you're wallowing in it.

We are starting to get a sense here for why boredom is a poor candidate for benign masochism. Unlike other masochistic pleasures, it doesn't capture our attention or our interest (indeed, it does the opposite). It doesn't lead to escape from the self. It might set up a contrast with future experience—something interesting is probably more interesting if it was preceded by a period of boredom—but apparently the contrast isn't so great as to make the choice worthwhile.

BOREDOM CAN HAVE *some* appeal—I said that it is rarely sought out, but I didn't say never. Take art. Elpidorou notes that much of it is boring: "The cetology sections of *Moby Dick* are boring. Satie's *Vexations*, if played in its entirety, is boring. Wagner's *Ring Cycle* is boring. And so is Warhol's *Empire*, William Basinski's *The Disintegration Loops*, much of slow cinema, and many second movements of symphonies." Elsewhere he observes that Susan Sontag, in a diary entry, felt the same, giving her own list: "Jasper Johns is boring. Beckett is boring, Robbe-Grillet is boring. Etc. Etc." And yet some claim to like these things. Perhaps they don't see them as boring. Or perhaps they recognize that even the greats mess up sometimes, and there is enough good going on in, say, *Moby Dick*, to make up for the duller chapters on whale anatomy. But also, for some, the boredom can be part of the appeal. Struggling through difficult texts can be an engaging challenge, and it can also be a mark of status. You enjoy your Stephen King and Dean Koontz; I'll just be sitting here with my Kierkegaard and Knausgaard. I wouldn't be the first to say that taking pleasure in difficult and boring literature can be yet another form of signaling.

Also, people have all sorts of motivations. Sometimes it's

just orneriness. When some philosopher or critic tries to define art, some smart-ass artist will create something that doesn't fit the definition, the subversion being the very point of the creation. Similarly, clever people might think of reasons to embrace boredom. One (fictional) example of this comes in Joseph Heller's *Catch-22*, where we're told that "Dunbar loved shooting skeet because he hated every minute of it and the time passed so slowly." Dunbar, it turns out, cultivates boredom and misery because he believes it extends his life, his subjective experience. After all, enjoyable and engaging experiences go by quickly, bringing one closer to death. His friend challenges him:

> "Well, maybe it is true," Clevinger conceded unwillingly in a subdued tone. "Maybe a long life does have to be filled with many unpleasant conditions if it's to seem long. But in that event, who wants one?"
>
> "I do," Dunbar told him.
>
> "Why?" Clevinger asked.
>
> "What else is there?"

3

AN UNACCOUNTABLE PLEASURE

The capacity to take pleasure in suffering is part of human nature. Masochistic appetites arise in every society and within every individual, though their precise form might vary. This is unique to our species.

Some scholars will cringe at this last assertion. They'll tell you about the long history of scientists making confident claims that only humans have language or the ability to reason about the future or the desire to be kind toward strangers . . . and then about other scientists going out in the field or doing laboratory experiments and finding that, no, actually, such capacities aren't unique at all. Other creatures have them, too, though always to a limited extent. These critics might go on to insist that a good Darwinian, one who appreciates evolution and common descent, will talk only about differences in degree, not kind.

I have some sympathy for this response. It's appalling to

hear some of my colleagues in the humanities talk about human sexuality and human sociality with an almost militant disinterest in our evolutionary history or the lives of other primates. Humans weren't created from scratch, and anyone who is seriously interested in how our minds work simply must have some appreciation for both the logic of evolutionary theory and the rich body of data on the mental lives of other animals. The novelist Ian McEwan, for instance, is saying something deep when he observes that the main themes of the nineteenth-century English novel match nicely with the lives of pygmy chimpanzees: "alliances made and broken, individuals rising while others fall, plots hatched, revenge, gratitude, injured pride, successful and unsuccessful courtship, bereavement and mourning."

But there's nothing wrong with claiming that certain traits are uniquely human. It's a misunderstanding of evolutionary theory to say that genuinely new organs or new capacities can't evolve. Obviously, they do. Species diverge over time; they take on their own special traits. Only an elephant has a trunk, for instance. And it would be mad to deny that there are qualitative differences, not just differences in degree, between cockroaches and people. The millions of years that separate us from other primates is ample time for novel capacities, including novel psychological capacities, to have come into being.

And even if you knew nothing about evolution, it's clear that there is a lot about humans that has no parallel anywhere else in the biosphere. We are the species with Broadway plays and moon landings and mathematics departments and protests against political oppression—and, on the other side of things, we are the species with torture chambers and concentration camps and nuclear weapons.

Now, all of this arises through communities of people over time. There are societies right now, each member of these societies fully human, that lack the rudiments of science and technology, where nobody reads or writes. And yet, individuals in these societies have the right mental capacities; any neurologically normal member of our species is equipped to engage in these wondrous human activities.

When scientists speculate on precisely what's so special about humans that gives rise to the world we live in, they'll tend to talk about our powerful social nature or our capacity for cultural learning or the power to form abstract categories or our gift of language. But what's often missing is what I see as the greatest gift of all—the imagination. Humans are blessed with the power to conjure up worlds that don't exist and might never exist. And this changes everything.

THIS CHAPTER IS about how our unique powers of imagination give rise to certain pleasures. But we likely evolved these powers for two reasons that have nothing to do with pleasure.

The first is dealing with other people. If you can imagine an alternative world, then you can see things through someone else's eyes, even if their sense of reality doesn't match your own. This makes possible perspective-taking, empathy, and much else.

Seeing the world through the eyes of others is essential to many acts of kindness. For me to respond to your worries and alleviate your fears, I need to understand your thoughts, even if I don't share them. (I might soothe a child who is terrified of a small dog, even if I'm not frightened in the slightest.) For me to successfully teach you, I have to imagine what it's like not to know something; only then can I get the new information into your head without either flummoxing you or boring you. Even activities as seemingly banal as buying a birthday gift or talking to a young child require the capacity—sometimes called social intelligence, emotional intelligence, theory of mind, or cognitive empathy—to imagine the world as seen and experienced through the eyes of another. Our altruism and kindness are grounded in the capacity to imagine the world as others see it.

But so is our cruelty and manipulation. Another name for this capacity to suss out the minds of others is "Machiavellian intelligence," and the name captures the dark side of this power. In order to lie, negotiate, seduce, charm, cheat, bamboozle, and trick, one also needs to appreciate that the world in your target's head won't match the one in yours. And so I might say to you about a man I have a grudge against, "Don't befriend him, because he is a liar and a cheat," when actually I know this is a lie. (I know he's the nicest guy in the world.) In order to pull this off, I have to appreciate that I am creating in your head a picture of a world that is both different from my own and different from the truth, and for me to proceed with this—"Who has he cheated?" you ask skeptically, and now I have to continue with my story and embellish it—I need to imagine the world as *you* now see it, hold in my head not just an accurate representation (for my own personal use) but also a different, false picture of the world. Lying is taxing, because of the problems of keeping two conflicting sets of books, but our extraordinary powers of imagination make it possible.

There are those who are especially good at this understanding of other minds—gifted teachers, seducers, psychologists and psychiatrists, and torturers. And there are those who find it hard. Some of us are awkward, self-centered, slow on the uptake, constantly puzzled by minds that don't match our own. Having serious problems with this sort of social reasoning is one of the hallmarks of autism, and it's one reason why the lives of individuals with this condition can be so difficult, even though they are often otherwise highly intelligent. Because they can't easily understand other minds, social interaction is a struggle.

We all struggled when we were young. There is debate over how much social imagination babies and young children have, but plainly they are pretty bad at imagining the world as seen by others. Young children are awful liars, because they can't grasp others' minds well enough to mess with them; they are the sort to deny eating the cake even when their mouths are smeared with chocolate. They are awful at hide-and-seek for the same reason, having little capacity to orchestrate the world so as to avoid being found by others. Nonetheless, as I've argued elsewhere, even young children are, relative to the adults of other species, surprisingly gifted at appreciating others' thoughts.

When psychologists talk about the function of conjuring up worlds that don't really exist, we tend to focus on such social factors. But the second function of this mental system is more general: planning for the future. Successfully making it through the world involves conjuring up multiple scenarios and seeing how each of them fares, like a chess computer generating multiple forking futures and evaluating each of them so as to plan the best move. If I complain to my boss, *this* will happen; if I speak to a lawyer, *this* will happen; if I do nothing, *this* will happen. Without the imaginative capacity to depict these nonexistent futures, and to mark them as unreal, we are trapped in the present; we just have to choose an option, see how it works out, learn from our experience, and try to do better next time. As the philosopher Anthony Nuttall put it, through the imagination "the human race has found a way, if not to abolish, then to defer and diminish the Darwinian treadmill of death. We send our hypotheses ahead, an expendable army, and watch them fall."

Just the human race? There is some hint of imaginative power in other critters. They can plan, in limited ways. They dream. Some monkeys and chimpanzees show signs of understanding other minds and perhaps even understanding when other members of their species are ignorant or mistaken, though the extent of their powers is the focus of a lot of debate.

But nobody doubts that the capacity of even our closest relatives to engage in social reasoning and planning for the future falters relative to that of humans, even human children. And there's no evidence that creatures other than us can consciously control their imaginations. Humans are unique in that we spend so much of our experiential life in the past, in possible futures, and in the minds of other people. Chimpanzees are trapped in the here and now.

OUR IMAGINATIONS ARE constantly humming along. When you add up the hours, the dominant pleasure of life is not being with friends and family, playing games, participating in sports, or having sex—though people do enjoy all of these things, or at least claim to when they fill out surveys. It's not anything we actually do or experience in the real world. Instead, the most common pleasures involve experiences that don't really exist, as when we read novels, go to movies, play video games, and daydream. They are pleasures of the imagination. This is how we spend most of our time—Netflix without the chill.

What something has evolved for and what something actually does are two separate things. Once we come to possess a capacity, we can use it for unintended purposes. When it comes to our imagination, we behave like a teenager who owns a powerful computer, dutifully sits in front of it after school to prepare for the AP calculus exam, . . . and then spends every free minute watching porn, playing *Call of Duty*, and gossiping with friends on social media.

So, yes, if you were to ask your genes to weigh in, they'd tell you to stop fantasizing and get to work on reproductively useful activities—eating and drinking and fornicating, establishing relationships, building shelter, raising children, and so on. But our genes should be used to this sort of rebellion by now. It's Evolutionary Biology 101, after all, that something can evolve for one purpose and be co-opted for another. Noses can hold up glasses, toenails can be painted so as to impress one's lover, and our thumbs are capable of lightning-fast texting, but none of this is what these body parts originally evolved for.

This process of co-option holds for certain psychological capacities as well. The capacity for rational thought has evolved to deal with the demands of existing in a complex and often zero-sum social and physical world, but it's now used to theorize about the origin of the universe, strategize our way through fantasy football, and engage in a billion other non-reproductively relevant tasks. We are the species that is clever enough (and self-destructive enough) to invent so-called supernormal stimuli that are far more potent than what our minds originally evolved to cope with. And so we gorge ourselves on family-size bags of M&Ms and Coke, basking in the pleasure generated through the overtriggering of a universal human love for sweets, something that began as a perfectly respectable adaptation in a world where we were at perpetual risk of starving to death—and where sweet fruit was a nutritious lifesaver—but which now kills us.

Or take sex. Our minds have evolved to respond to certain experiences with sexual interest because this was an important step toward having sexual intercourse, which is an excellent activity to engage in from a Darwinian perspective. But we can be tricked, or allow ourselves to be tricked, by experiences that aren't actually generated by the real thing. This isn't just a human vulnerability. Male rhesus monkeys will "pay" (by giving up fruit juice) to get the chance to see pictures of female monkey hindquarters. In the 1950s, researchers were interested in sexual behavior in male turkeys, and they found that they would mount a lifelike model of a female turkey, and would also do so for a model without a tail or feet or wings; indeed, male turkeys were fully aroused by a female turkey head on a stick.

And people do pretty much the same thing, entranced by patterns of light on a two-dimensional screen. Unlike the turkey, though, we fully appreciate the difference between the surrogate and reality, between the representation and the real thing. Also unlike the turkey, we can create our own representations for the pleasure of ourselves and others. Presumably, a long time ago in the history of our species, one of our more creative and pervy ancestors got turned on when scratching figures in the sand or painting on a cave wall or molding a statue just so, and pornography was born.

The co-opting of evolved systems can get us into trouble. The capacity to create foods with immense amounts of fat and sugar and salt has not been, on the whole, a good thing for the health of us moderns. Some of us apparently get more pleasure from video games and streaming video than from interacting with real people. And for some, pornography has replaced the far more reproductively useful (though far more complicated) act of actual sex. All of this can interfere with the rest of life.

WHAT, PRECISELY, MAKES the imagination so much fun?

To some extent, it's because of the phenomenon illustrated by the turkey head on a stick. We often respond to the imagined in the same way as we respond to the real. Philosophers have long known this, though instead of lust, they usually give the example of fear. Montague wrote that if you put a wise man on the edge of a precipice, "he must shudder like a child." David Hume imagined someone hung out of a high tower in a cage of iron; though he knows he's perfectly secure, still he "cannot forbear trembling."

Our minds are in part indifferent to the contrast between experiences in the real world and experiences that are similar to the real world that arise through the imagination. And so we can create and consume surrogates of pleasant real-world experiences, know that they aren't the real thing, but nonetheless experience the pleasure that they would elicit if they were real. Even babies and young children can appreciate this—think of peekaboo, or of a parent sending a child aloft while vrooming like a rocket.

This is the simplest sort of imaginative pleasure—the simulation of worlds that satisfy real-world appetites. If we like real sex, we fantasize about sex. If we are lonely, we surround ourselves, in our imaginations, with interesting people, as when we consume books, movies, and television, but also when we conjure up imagined conversational companions. If we aspire to a full life, with love, adventure, and triumph, we can partially scratch the itch by putting ourselves into the shoes of others, real or imagined, who are fulfilling that life. We can do all this by ourselves, just by closing our eyes and creating new worlds, but often we can immerse ourselves in worlds created by others who are more creative and skilled than we are, and this gets us to fictions ranging from the Avengers superhero series to Shakespearean comedies.

The imaginings of others are of course less personal than the ones we create for ourselves—I would be very lucky indeed if Spielberg or Tolstoy or some California pornography studio were able to create just the imaginary world that best connected to my desires. But others are typically better than I am at dialogue, plotting, sound effects, and so on. Watching *Godzilla* on the big screen is far more vivid than closing my eyes and trying to conjure him up; the creators of *Seinfeld* and *Fleabag* create much better dialogue than I can.

One special form of imaginative experience involves memory—replaying reconstructions of past events. All animals have memories, but as far as we know, only humans can willingly recall experiences from the past and savor them. This is going to be important later on, as we'll see that our decisions about how to live our lives aren't just based on the experience we'll have at the time, but also on our sense of what it will be like to look back at the experience in the future.

Then there is anticipation. We discussed this in the context of the contrast theory of benign masochism, but its appeal is more general. One ingenious study asked this: Suppose you could kiss your favorite movie star on the lips, willingly and consensually on their part, with no repercussions. You can do this just once, but it can be anytime in the future that you want. How much would you pay to get to the kiss right away, as opposed to after a delay? When *should* you want it the most?

If you were asked this in your first exam of an introductory class in economics, the answer your professor would probably look for is: *Right now*. Everyone knows that a dollar now is worth more than a dollar later, and the same argument applies here: Who knows what could happen in the future? You could die; the movie star could die; you could lose interest in kissing; the mysterious offer might go away. And from a psychological view, "Right now" is again the best answer. Our minds have been adapted to appreciate the economists' insight about the value of a bird in the hand, and so we are temporally greedy; we often choose one marshmallow now rather than two marshmallows in the future.

But what people actually prefer is to get the kiss after a delay of a few days—they will pay more to get the kiss in three days than to get it in three hours or twenty-four hours. People apparently like to savor the idea of a pleasant experience, holding back a bit, saving the best for later. (But not too much later—they pay more for a kiss in three days than in one year or ten years.)

But why not get the kiss right away, and then follow it up with the willful imaginative reconstruction of the past, savoring the memory? I don't know. Maybe anticipation of the future is more pleasant than memory of the past? Or maybe memory and anticipation are different pleasures, so people want a taste of both? Regardless, anticipation is an important source of imaginative pleasure.

—

We've talked so far about the use of imagination as a substitute for certain real-world pleasures—Reality Lite. But some of our imaginative pleasures seem to be the perfect opposite of what you'd expect from such an account. When we can experience whatever worlds we damn well please, we often conjure up and seek out terrible ones, replete with all sorts of suffering. Milton had Satan say,

> The mind is its own place, and in it self
> Can make a Heav'n of Hell, a Hell of Heav'n.

And we often choose to create hell. We choose to experience suffering in two related ways—we can take delight in the suffering of others, often imaginary others, and we can enjoy the suffering that we experience directly. Anna Karenina kills herself out of misery, and this makes us miserable, too, and we love it.

I talked about horror movies earlier in this book as an example of the pleasures we get from vicarious suffering, but a glance at Shakespeare plays would make the same point—you get murder, torture, rape, the whole shebang. *Titus Andronicus* involves a scene in which rapists, to ensure the victim's silence, have her tongue cut off and her hands amputated. The conclusion has a mother tricked into eating a pie made from the bodies of her sons. (Which was the inspiration for a similar scene in the HBO series *Game of Thrones*.)

Or consider the glories of the Colosseum in ancient Rome. A typical day started with exhibitions of animals, including exotic animals—and then their slaughter. Lunch and early afternoon were spent on executions, carried out by having the convicted criminal killed through forced combat, being burned alive, or being eaten by animals such as lions. Sometimes the executions were done as reenactments of myths. For instance, the story of Icarus flying too close to the sun was depicted by having convicts pushed off a high tower.

Later came the gladiator matches. In his *Confessions*, Augustine tells of Alypius, who was dragged to the spectacles by his law school friends; he was disgusted by it and kept his eyes closed. But then the roars of the crowd got to him and he took a peek: "When he saw the blood, he drank in the savagery and did not turn away but fixed his gaze on it. Unaware of what he was doing, he devoured the mayhem and was delighted by the wicked content and drunk on its cruel pleasure. . . . He looked, he shouted, he was fired up, and he carried away with him the madness that would goad him to return." The classicist Garrett Fagan, who recounts this story in his *The Lure of the Arena*, wonders if this was Augustine describing his own experiences, too ashamed to attribute them to himself.

Alypius and Augustine are men, and it's true that these entertainments, including modern horror movies, appeal more to men. But the sex difference isn't as big as you might expect. One study of a thousand people asked them to rate how much they liked horror movies on a scale from 1 to 5. Men averaged 3.5, and women averaged 3.3—a statistically real difference, but hardly a strong one.

Horror movies have a lowbrow reputation, but one finds similar features in more elite forms of pleasure. Take Broadway plays. In 2008, there was *Blasted*, which, according to enthusiastic coverage by the *New York Times*, involved rape and eyeball eating and audience members passing out. And, a couple of years ago, there was a production of *1984* that was so graphic it was reported to cause audience members to "faint and vomit."

Some prefer to indulge in sadness instead. There is a popular television show called *This Is Us* that apparently makes everyone cry. I read a magazine article titled "Your Weekly Cry-Fest Over 'This Is Us' Has Surprising Health Benefits" and learned from an online review that this show "hits us right in the feels." In certain regards, sadness might be more pervasive in our culture than fear. There are, to my knowledge, no horror songs, but there is plenty of sad music.*

I've been talking so far about the consumption of negative experiences created by others, but we often create negative experiences in our own heads. We are blessed with the power to think of whatever we choose, and yet we often choose to think about what makes us sad. This was explored in a paper by Matthew Killingsworth and Daniel Gilbert called "A Wandering Mind Is an Unhappy Mind." They used an "experience sampling" method—an iPhone app that bothered the subjects at random parts of the day. When the phone went off, they had to answer a happiness question ("How are you feeling right now?"), an activity question ("What are you doing right now?," with a menu of twenty-two common activities), and a mind-wandering question ("Are you thinking of something other than what you're currently doing?").

They found that people's minds wandered *a lot*, just under half the time, and the timing of mind-wandering wasn't influenced by how happy they were or what they were doing. Less than half of the mind-wandering experiences were positive, and more than a quarter were reported as unpleasant. On the

---

* On the other hand, there are genres, such as heavy metal and screamo that, while they don't exactly elicit fear, do often lead to agitation and anxiety. (Thanks to Alexa Sacchi for discussion of this point.)

whole, people were less happy when they were mind-wandering than when they were not.

The lure of the aversive isn't limited to adults. In many ways, children are fragile, easily frightened, not as capable as adults of distinguishing the imaginary from the real. Still, parents are often shocked by their appetite for violence and gruesomeness. Jonathan Gottschall observes, "The land of make-believe is less like heaven and more like hell." Consider this transcript of a pretty typical preschool play session, between a three-year-old (Marni) and a four-year-old (Lamar), originally recorded by the child anthropologist Vivian Paley:

Teacher: Where's the baby, Marni? The crib is very empty.
Marni: The baby went to someplace. Someone is crying. . . . Lamar, did you see my baby?
Lamar [at sand table]: Yeah she's is in a dark forest. It's dangerous in there. You better let me go. It's down in this hole I'm making.
Marni: Are you the daddy? Bring me my baby, Lamar. Oh, good for you, you finded her.
Teacher: Was she in the dark forest?
Marni: Where was she, Lamar? Don't tell me in a hole. No, not in a hole, not my baby.

Our delight in the imagined negative has long fascinated philosophers. David Hume provides the classic framing of the puzzle (which also gives us the title of this chapter):

> It seems an unaccountable pleasure which the spectators of a well-written tragedy receive from sorrow,

> terror, anxiety, and other passions that are in themselves disagreeable and uneasy. The more they are touched and affected, the more are they delighted with the spectacle. . . . They are pleased in proportion as they are afflicted, and never are so happy as when they employ tears, sobs, and cries to give vent to their sorrow, and relieve their heart, swoln with the tenderest sympathy and compassion.

Implicit in how Hume sees the problem is a certain assumption. Some psychologists think that those who enjoy horror movies aren't actually scared and those who enjoy tragedies aren't actually saddened. Or they think that these negative emotions are the costs that one is willing to pay for future pleasures—what you have to put up with to get to the good parts. By contrast, Hume thinks that the pleasure of such experiences is in proportion to one's experience of anxiety, sorrow, and the like. To put it in modern terms, for Hume, the negative emotions are features, not bugs.

Hume is right. After all, people who like horror movies are up-front about enjoying them just because they are scary. This response, from a subject in a psychology study, is typical: "It may seem masochistic, but the more scared I feel watching a horror movie, the more I enjoy it!" Similarly, if you told someone that the sadness of *This Is Us* was just a part of the show that you had to put up with to get to the good parts, they would turn their tear-stained face at you in confusion. They are in it for the feels! My son Zachary, at around age four, was watching a cartoon one day with a violent chase scene, and he began to get agitated and started to sniffle, so I said, "No worries, Zach," and reached for the remote control. He turned and yelled at me, still crying, "Don't turn it off!"

And there is research supporting this, done by Eduardo Andrade and Joel Cohen. They tested two types of people—horror fans, who claimed to watch horror movies at least once a month, and horror avoiders, who hardly ever watched such movies. They got both groups to view some frightening clips from *The Exorcist* (possessed twelve-year-old spews green vomit, shouts obscenities, and swivels her head like a dreidel) and *Salem's Lot* (small town invaded by vampires, adorable child gets bitten, unhappy ending).

After watching the scenes, everyone filled out surveys to describe their experiences. Both the horror fans and the horror avoiders claimed to experience negative emotions. But only the fans also claimed to experience positive emotions. They weren't insensitive to the fear and anxiety, then—they experienced it plenty—but they, unlike the others, took pleasure in it.

In a subsequent study, subjects got to report their reaction to the movie as they were watching it, using a mouse to record the extent to which they felt, on a scale from 1 to 5, *afraid*, *scared*, *alarmed*, *happy*, *joyful*, or *glad*. For those who didn't like horror movies, responses of fear were negatively associated with responses of happiness, but for horror fans it was the opposite: fear and happiness went up and down together.

A related study showed people thirty-eight movies that each contained a one- to two-minute clip of a specific sort of scene—one where a character learns that someone they love has died (as in *Mystic River*, when the character played by Sean Penn learns that his nineteen-year-old daughter was murdered.) The more saddened people were by the scene, the more they wanted to see the rest of the movie.

HUME ALSO GOT two things wrong. The first is subtle, having to do with the phrase "to give vent to their sorrow." This focus on the expression of sadness is a version of the catharsis theory developed by Aristotle and extended by Freud, in which there is said to be a sort of purging process—when we experience fear and anxiety and sadness, these emotions are released, and we then feel calm and purified. This is a popular theory of why we enjoy aversive fictions: we do it for the positive payoff at the end, a sort of cognitive enema.

This isn't ridiculous. There are those who claim to feel relaxed after a good cry. But in general, it's just false that negative emotional experiences have a purging effect. Many walk out of a great horror movie shaken up and maybe, for the next little while, keep the lights on at night. In a recent survey of horror movie fans, most said that they were more scared after the movie ended; only one in twenty said they were less scared. Of all dead psychological theories, catharsis is the deadest.

The bigger problem, though, is with how Hume framed the puzzle. For him, this is a puzzle about certain sorts of fictions. This is why it's called "the paradox of tragedy," after all.

This is also a popular theory, with a long and illustrious history. Shifting the art form a bit, Aristotle wrote that "the sight of certain things gives us pain, but we enjoy looking at the most exact imitations of them, whether the forms of animals which we greatly despise or of corpses." Certain imitations—but not certain things. In *The Lives of the Poets*, Samuel Johnson wrote, "The delight of tragedy proceeds from our consciousness of fiction; if we thought murders and treasons real, they would please no more." Fiction—but not reality.

I think this is mistaken. Shakespeare's tragedies depict precisely those events we find most interesting in the real world—interactions involving sex, status, family, and betrayal. I remember watching the actual O. J. Simpson trial and how enthralling it was, and certainly its reality didn't diminish in the slightest its appeal or the appeal of the documentaries and dramatizations that followed it. The death of Princess Diana was so affecting because it was real. When a memoir is discovered to be fictional, its sales go down, not up. It takes nothing from a tragic story to hear that it really happened.

Or consider Aristotle's claim about the disgusting reality—which he says "gives us pain"—versus the pleasing imitation. Aristotle would reconsider this if he had been in the car with me and my sons a little while ago, rushing to get to a movie, and there was a bad accident on I-95, which meant a lineup of cars grew as people slowed down to rubberneck. We all sat there, annoyed at the ghoulishness of some people, and then we finally got to the scene of the accident and also slowed to get a good look—*Oh my God, look at all the broken glass. Is that blood?* Once, when driving to work, I noticed on a street corner one of those old coin-operated newspaper boxes and could see the headline—GRUESOME DETAILS—and I made a note to myself to read the story online later, because really, what better details are there? Or take Plato's *Republic*, where Socrates tells the story of Leontius, who is walking in Athens and sees a pile of corpses, men who had just been executed. He wants to look at them but turns away, struggles, at war with himself, and finally runs to the corpses and says to his eyes, "Look for yourselves, you evil wretches, take your fill of the beautiful sight!"

HUME IS WRONG, then, when he assumes that our appetite for negative experiences exists only when we think the experiences are unreal. Still, the imagination is a particularly good realm through which to experience the negative. This is because the imagination is relatively safe.

After all, it might be (for reasons we haven't yet explained) exhilarating to savor the feeling of an ax murderer stalking a small town, but, for obvious reasons, I really don't want an ax murderer in my own small town. As Edmund Burke put it, "Terror is a passion which always produces delight when it does not press too close." Similarly, if we eavesdrop on a real conversation, we run the risk of being caught and embarrassed; spying on people having sex through their bedroom window is even riskier; and actual sex might lead to pregnancy, disease, or physical and emotional harm. In general, fiction—and here there is a commonality with nonfictional outlets like history, journalism, documentaries, and the like—allows us to get our kicks in ways that are, relatively speaking, risk-free. Nobody ever froze to death watching *Everest*.

Fiction is safe in a different way as well, in that it allows for control of what kind of aversive experience one is going to get. This is particularly true since about a decade ago, with technology offering us virtually unlimited choice. A friend of mine recently went on social media and asked her many followers for a recommendation for a television show that could help her fall asleep; she wrote, "I like feminine-coded television, dislike violence or disturbing imagery, and nothing is too lowbrow." We have the luxury of being picky.

Some people take this desire for control to the extreme. A former student of mine, Jennifer Barnes, studies, among other things, romantic fiction, and she points out that there are book series that cater to people who have very specific tastes. They have strict plot conventions. A common theme, for instance, is the *Nanny and the Billionaire.* You can guess how this goes; there are strict rules as to precisely what happens, and the reader can consume book after book in which the plots are virtually identical. Movie sequels that show the original plot over and over again, with some minor modifications—think of *Die Hard 2* or *Speed 2*—scratch a similar itch.

If you *really* want to avoid surprises, the best trick is to read a story repeatedly. Heraclitus said that you cannot step into the same river twice, but you can come close if the river is fictional, and the safe pleasure of a repeated story is particularly compelling to young children.

To appreciate negative stories, in fiction and in reality, you need to have a certain distance—not too close or too far. You have to immerse yourself deeply enough so as to worry and obsess and fear; you have to care about Anna Karenina, Little Dorrit, Ned Stark, and Dobby the house-elf, but, particularly for fictions where things go wrong, you also need to appreciate that none of these characters are real, so that empathy and anguish and concern won't override the pleasure. There is a Goldilocks principle here; one might call it a sweet spot.

Children are poor at distancing, so they have less tolerance for horror than adults. There is an elegant experiment that makes this point, in which psychologists asked four- to six-year-old children to pretend that there is a monster in a box.

Children then often refused to put their fingers in the box. It wasn't that they were confused—they knew it was pretend—it was that they had a problem segregating the imagined and the real. You don't need a research study to understand that horror movies can give children nightmares.

And, indeed, I'd bet that the same experiment would work for adults in a more subtle way. If you showed adults two boxes, one of them the monster box ("Pretend that the monster in the box loves to eat fingers, chews them right off!"), the other a normal box ("Check out this empty box"), and then asked them to put their hands into both boxes, I predict they would hesitate, maybe for just a fraction of a second, before putting their hands into the monster box. I believe this in part because of Paul Rozin's discoveries that people often refuse to drink soup from a brand-new bedpan, eat fudge shaped like feces, or put an empty gun to their head and pull the trigger. As Tamar Gendler points out, the mind works on two tracks. We know, consciously, that the bedpan is clean, the fudge is fudge, the gun is empty, and yet we can't help blurring the imagined and reality; our minds scream, "Dangerous object! Stay away!"

WE'VE TALKED A lot about the proper framing of Hume's problem, but we haven't yet tried to answer the question of why we like such aversive experiences at all. What's the appeal of fictional suffering?

To start, consider again one theory of benign masochism discussed in the last chapter. The idea is that we get pleasure through contrast, by creating situations where the release from unpleasantness is its own source of pleasure. Think about slowly sinking into a painfully hot bath and then gradually adapting to the temperature, experiencing the soothing contrast with the initial pain. Or the burn of hot curry balanced by cool beer. Or the pain of rigorous exercise and how good it feels when it's over.

Many stories work the same way—initial pain and difficulty set the stage for triumph at the end. Earlier I gave the example of revenge tales, where we squirm through some injustice and are later rewarded with the satisfaction of vengeance. This is nicely encapsulated by the tagline from the remake of *Death Wish*—"They came for his family. Now he's coming for them"—where the first sentence describes the negative, the second the positive. Bad-then-good is the contour of children's stories such as *The Little Engine That Could*, where the initial struggles ("I think I can, I think I can") make the engine's victory at the end ("I thought I could! I thought I could!") all the sweeter.

This sort of structure is common. David Robinson analyzed a database of 112,000 plots of stories downloaded from Wikipedia, from books, movies, video games, TV shows, and so on. He did a range of analyses on this database, including "sentiment analysis," which is a method of looking at the positivity and negativity of chunks of text. The overall pattern is that stories begin on a high point and then gradually descend, becoming more and more negative until, just before the ending, they rise sharply in positivity. As Robinson puts it, "If we had to summarize the *average* story that humans tell, it would go something like 'Things get worse and worse until at the last minute they get better.'"

Is it a problem with this theory that we get pleasure from an aversive story even during the unpleasant parts? After all, we know from the studies described above (and from common sense) that the tears and the scares are part of the appeal of the stories, not something we suffer through for the pleasure at the end.

Not necessarily. As we discussed earlier, in the context of benign masochism more generally, there is the pleasure of anticipation. We know that this is a story and we know how stories work, or at least typically work, so our initial struggles are mixed with anticipatory pleasure. Even during the "They came for his family" part, we know that the "He's coming for them" will soon arrive, and that changes our experience. This is one critical regard in which fiction is different from reality. Unless one has strong beliefs about a benevolent and loving God, life has no screenwriter and no director. And so, when we suffer through bad events, we can't be confident that things will work out in the end.

I am skeptical about this bad-then-good explanation, though. First, not all stories have this structure. There are many taxonomies of stories out there, and it turns out that there are other ways to balance the good and the bad in an appealing narrative. Making this point, another analysis chugged through thousands of works of fiction, analyzing their emotional content as the stories progressed, and found that the stories fell into six main categories, only some of which end on a happy note:

1. Rags to Riches (rise)
2. Riches to Rags (fall)
3. Man in a Hole (fall then rise)
4. Icarus (rise then fall)
5. Cinderella (rise then fall then rise)
6. Oedipus (fall then rise then fall)

This variety holds for aversive fictions as well. Yes, many horror movies end with the monster being killed, but many don't. And isn't an unhappy ending part of the very definition of a tragedy? Any good theory of aversive fictions can't simply rely on this sort of payoff model as an explanation for our pleasure.

Indeed, I don't think that bad-then-good is the right explanation even for Man in a Hole stories. Consider watching some bad events, then some unrelated bad events, then some more bad events, and then some unrelated positive events. There, you now have a bad-worse-good contour. Things get worse and worse until at the last minute they get better. But I doubt it would be appealing.

Rather, Man in a Hole is appealing because it tells a certain sort of story. In particular, it is about the overcoming of an obstacle. The value of such a story is old news. It is at the core of Joseph Campbell's theory of stories, what he calls the hero's journey. It is part of all seven of the "seven basic plots" outlined by Christopher Booker. It is central to many cognitive science approaches to literary universals, especially the work of Patrick Colm Hogan. It's in Aristotle's *Poetics*, as part of the instructions on how to create a tragedy. And it's, almost literally, Screenwriting 101. In an online class (part of the MasterClass series) on the topic, the screenwriter Aaron Sorkin says that the most foundational and basic advice for story construction is to *present a formidable obstacle*.

In a lovely paper called "Suspense in the Absence of Uncertainty," Richard Gerrig points out that suspense can be created even if one knows the outcome—the election of George Washington as president, say, or the successful creation of the atomic bomb by the United States in World War II—so long as there is uncertainty about how the obstacles are dealt with. It is this surmounting of obstacles that can pull us in; they're what give the opportunity of pleasure.

This focus of obstacles explains why happy endings are optional. There is a pleasure to having the hero succeed, but this isn't essential. Man in a Hole, yes, but it's not necessary that he ultimately gets out. Two of the greatest sports movies of all time are (in my humble opinion) *Rocky* and *Friday Night Lights*, and neither ends with victory. Nor do many great war movies. Nor do any of the John Wick movies (so far). Happy endings are overrated.

Also, this focus on obstacles explains why so many of the stories that we enjoy have aversive elements. Obstacles get in the way of what you want. Now, the obstacle in a story might be mild or even amusing, as in a romantic comedy where, say, a couple is trying to spend time together despite the machinations of intrusive parents, or a children's book where the train really wants to make it up the hill. But the difference between this and a story in which a man tries to escape after being buried alive is merely a matter of degree. In all cases, there is some level of anxiety and stress. Without this, the stories would have no dramatic tension and would be boring.

Finally, a focus on obstacles makes clear how the attraction of aversive fiction connects to what draws us in real life. In our actual lives, we seek out projects with difficulty and struggle, ones that involve surmounting obstacles. This is a large part of what gives life meaning. But this is a topic for the second half of this book.

THE SECOND THEORY of why we enjoy aversive fictions involves play.

Children, left to themselves, choose to play. They pretend to be airplanes or to have tea parties or make war, or they just grapple and race and knock one another down. Other creatures, like dogs and cats, also play, sometimes violently, sometimes with unfortunate smaller living things. And adults play as well (though we don't usually call it this), in gyms and dojos and stadiums and arenas.

One common theory is that play reflects an evolved motivation to practice. Fighting is the best example of this. Being good at fighting is useful, and one way to get better at fighting is to get experience fighting. But getting into real fights is dangerous—you can get killed or seriously injured, or you can kill or injure another. Evolution has come up with an ingenious solution to this problem: we can play at fighting. We can find someone we like and trust and go through the moves of fighting and get better at it—but with various constraints to reduce the risk of harm. Modern humans can articulate rules in our play—no biting, no hitting below the belt, no kicking a man when he's down, tapping out, saying "uncle!"—and can also use special tools like gloves, helmets, and mouthguards.

The more we do something, the better we get at it, so we are drawn to immerse ourselves, in a safe way, in challenging physical, social, and emotional situations. Want to get better at flying and landing a plane? You can use a real plane, but it's safer and smarter to log hundreds of hours on a flight simulator. Well, imagination is a flight simulator—and you don't always program a simulator for a smooth flight. You often use it to prepare for trouble.

Just as physical play fighting involves thrusting oneself into a situation that would be dangerous if real, our imaginative play often takes us into situations that include elements that would be unpleasant, sometimes terrible, if we were really to experience them. The idea here is best summed up by Stephen

King: "We make up imaginary horrors to help us deal with real ones." It's "the tough mind's way of coping with terrible problems." We are drawn to tragedy and horror, then, because they are creative representations of worst-case scenarios, such as being attacked by strangers, being betrayed by friends, or experiencing the deaths of those we love.

There is a well-known critique of this view by Jerry Fodor. He quotes Steven Pinker defending the adaptive utility of ruminating about fictional worlds: "Fictional narratives supply us with a mental catalogue of the fatal conundrums we might face someday and the outcomes of strategies we could deploy in them. What are the options if I were to suspect that my uncle killed my father, took his position, and married my mother?" Fodor responds:

> Good question. Or what if it turns out that, having just used the ring that I got by kidnapping a dwarf to pay off the giants who built me my new castle, I should discover that it is the very ring that I need in order to continue to be immortal and rule the world? It's important to think out the options betimes, because a thing like that could happen to anyone and you can never have too much insurance.

At the risk of explaining a joke, Fodor's point is that the situations in fiction are often a poor match for the problems we really have to cope with. If so, then the practice theory is dead in the water.

But is he right? Actually, the themes of aversive fiction seem to be precisely those that are most relevant and most worrying. I'll repeat the catalog from Ian McEwan I mentioned earlier: "alliances made and broken, individuals rising while others fall, plots hatched, revenge, gratitude, injured pride, successful and unsuccessful courtship, bereavement and mourning." This is what fiction is all about, and isn't that precisely what we need to worry about in real life?

But what about magic rings? Fodor is correct that this is not very realistic, but it turns out that universal and relevant themes can be expressed in unusual and fantastical ways. Fodor was a huge opera buff—his example was a summary of the plot of Wagner's *Das Rheingold*—and I'm totally out of my depth here. But I do know my horror movies. Are the fantastic situations in these a counterexample to the practice theory? Well, it certainly would be silly to think we need to prepare for a zombie apocalypse. But this would be taking things too literally—the topic of zombie movies is never zombies. Rather, such movies are fantastical renditions of the very relevant worry of what would you do if society breaks down and the world goes to hell. (Almost without exception, the real danger in zombie films isn't zombies; it's other people.)

Indeed, some stylization is just what you would expect if fantasy is geared toward practice, because often practice strips away extraneous factors. It's not true that you have to practice at the same thing you want to get good at. Boxers, for instance, spend time working at the speed bag. This doesn't look anything like boxing in the ring, and as a simulation it's incomplete (the bag doesn't hit back). And yet it is useful practice for the real thing.

I said earlier that our use of the imagination as a tool for pleasure was an evolutionary accident, not an adaptation.

But this particular instance is an exception. Imagination-as-practice is an adaptationist theory, positing that our appetite for aversive fictions stems in part from our desire for play and practice, which in turn exists because of the benefits that having this desire had for our ancestors in the past. Those who could imagine bad scenarios and plan them in advance outlived and outreproduced those who could not.

So far we've been talking about the adaptive value of negative fantasies. What about pleasurable thoughts? There is a large research program by Gabriele Oettingen and her colleagues suggesting that positive fantasies can often be bad for you. One study looked at patients who were about to undergo hip replacement surgery the next day and asked them to simply imagine what it would be like for them two weeks later, when they had to do various activities like walking out to get a newspaper. The more positive the imagined experiences were, the poorer their later recovery was. In another study, college subjects with a crush on another student were asked to imagine future scenarios in which they got to know the crush better. The more positive the fantasies, the less likely the students were, months later, to have actually gotten together with that person. In other studies, positive fantasies about success in a course predicted lower grades, and positive fantasies of getting a job were related to not getting a job later on, and lower salaries when subjects did get one.

It's not clear exactly why positive fantasies—as opposed to positive *expectations*, which don't have these negative consequences—are bad for you, but one theory of Oettingen's is that they distract you from the actual goals you want to achieve. They serve as substitutes: If you get enough pleasure from the fantasy, you don't have to put so much energy into the actual pursuit. Thinking about failure and difficulty plainly doesn't have the same problem.

THUS FAR WE'VE discussed two explanations for the appeal of aversive fictions. First, they capture our interest in obstacles, reflecting what interests us most in the real world. And second, they serve as a form of imaginative play, allowing us to explore, in a safe way, dangerous and difficult situations.

There is a third explanation that I will get to, but before doing so, I want to talk about a specific sort of imaginative pleasure—though I'll admit ahead of time that I don't fully understand it.

We know a lot more about sexual fantasy than ever before. This is due to big data, and in particular to large-scale analyses of what people search for on pornographic websites. If you ask people to tell you about their sexual fantasies, they might lie or be too embarrassed to say. But the pornography they choose to watch is a fairly direct indicator of what turns them on. There are several sources of data here, but the biggest is from billions of searches on the porn site Pornhub. This data tells us not just what people like to watch but also what differences exist between young and old, gay and straight, men and women. (It turns out that porn sites use Google Analytics to determine their visitors' personal characteristics with high accuracy—who knew?)

When you look at the top searches, much of it is what you expect. Most of the searches are for physical features and body parts and sex acts that pretty much overlap with those that people would like to look at or experience in real life. This is just what you would expect under a Reality Lite theory, where imagination is a substitute for actual experience.

Then there are puzzles. For instance, cartoon pornography is quite popular. Seth Stephens-Davidowitz tentatively suggests that this reflects an obsession with childhood along Freudian lines. In support of this, he also notes that "babysitter" is a keyword commonly searched for by men. (Though an alternative for the babysitter fantasy is that these are young women whom some of the men who consume pornography might find themselves in close contact with, and so they are a natural source of fantasy.)

Then there is incest. Of the top one hundred searches on Pornhub by men, sixteen were incest themed at the time of Stephens-Davidowitz's study. The most common topic of these searches involved mothers and sons. For women, incest comprised nine out of the top one hundred searches, and the most common involved fathers and daughters.

Does this reflect an appetite for incest in the real world? I doubt it. Putting aside predation by older men, often stepfathers, there isn't much evidence for strong sexual attraction toward siblings, parents, or children. The biological constraint on close-family incest is a strong one. There are exceptions to everything, but, as Steven Pinker notes, "teenage brothers and sisters do not sneak off for trysts in parks and the back seats of cars."

What's going on, I think, is something a bit different. Imaginative pleasures are safe, but safety carries the risk of boredom. If someone waved a handgun in front of me in my office, it would be terrifying, but seeing someone wave a gun in a movie is ho-hum. We are used to it. To compensate, film violence can get very violent indeed. The same habituation takes place for pornography. There are teenage virgins who would swoon if they got kissed on the mouth by someone they found attractive but who have consumed so much pornography that they regularly seek out extreme depictions that I don't want to talk about in what I like to think of as a family-friendly book.

I suspect, then, that the popularity of incest porn might reflect an interest in the taboo, the shocking, and the inappropriate, mostly by people who have become jaded by more vanilla pornographic scenes. (Note also that many of the incestuous themes involve step-relatives, which is taboo but not as wrong.) A similar issue might arise in the interest in looking at leaked sex tapes, revenge porn, hidden cameras—all involving filming or distributing video of people without their consent or knowledge. The immorality here might block some from viewing them but might draw in others, just because it is forbidden.

Consider now a more disturbing sexual taste. In Stephens-Davidowitz's study, fully one-quarter of female searches in straight porn emphasize suffering—physical and psychological—with search terms including words such as "brutal" and "painful." (Five percent of searches were for "rape" or "forced sex," even though these are banned on Pornhub.) Despite the fact that men are, in the actual world, far more violent and more likely to commit sexual assault, these searches were at least twice as common among women than among men.

It's not just Pornhub. There is now a range of studies that ask men and women about what they fantasize about, with open-ended questions, interviews, and checklists. And several studies find that a large proportion of women (somewhere between 30 percent and 60 percent) claim to have rape fantasies, with about one-third of these women claiming that these are preferred or frequent fantasies. Because these sorts of fantasies can be embarrassing to admit to, even on an anonymous survey, such numbers are likely to be low estimates, not high ones.

So why the fantasy? One theory is that it might be a way to imagine sexual pleasure without the stigma or guilt of choosing to engage in sex. But this makes the prediction that women who engage in such fantasies would be more prone to be ashamed and reticent about sex, whereas the opposite seems to be true—such fantasies are associated with high diversity of sexual experience and more range of sexual fantasy.

What about the play-as-practice idea I just sketched out? Women are more vulnerable to sexual assault—it is something deadly serious that they have to contend with—so perhaps they prepare for it through fantasy.

But this theory fails to explain the phenomenology of the rape fantasy. Women *do* think a lot about being sexually assaulted, but they usually think of it in the same way that everyone thinks about being mugged or robbed or having one's child kidnapped, mulling such scenarios over in an unhappy way, thinking about how to defend against them. One typically doesn't get sexual pleasure from mulling over the worst-case scenario; nobody gets aroused thinking about having their credit cards stolen. Sexual fantasy is different; it is pleasant and arousing. So there is a poor mismatch here with practice theory.

I don't think anybody knows for sure what's going on here. My only speculation is that, while such fantasies are described as rape fantasies, and in a literal sense they are, when you look at them in detail, they have little connection to real thoughts of sexual assault. Deep dives into descriptions of these fantasies find that they tend to be stylized and unrealistic: The assailant is typically highly attractive; there is little or no physical pain, fear, or disgust; and the experience is pleasurable. They are not renditions of what it would be like to be raped; rather, they are versions of BDSM fantasies, mixed with other things—such as the notion of being so attractive that others lose control around you—and should be understood in the same way.

THE THIRD AND final explanation we'll entertain for the appeal of aversive fiction has to do with morality. We are fascinated by good and evil, and the right sorts of stories appeal to our moral natures.

A moral story requires unpleasantness. At minimum, to see good triumph, you have to witness bad things. In some stories, it might just be suffering without evil—a baby trapped in a well, say, who gets saved by rescuers. In other stories, there is suffering that is the product of evil—the damsel in distress, tied to the tracks by the mustachioed degenerate, rescued by our hero. In both cases, we see the same thing discussed earlier: a pleasure in witnessing obstacles overcome. But the presence of evil adds something special—it raises the possibility of payback.

When we think about the appeal of revenge, there is a temptation to think about lowbrow entertainments like the lesser Clint Eastwood films. And it's true that revenge tales—and, more generally, good-versus-evil conflicts—are often depicted in an extreme and exaggerated way. Comic books are one example of this. David Pizarro and Roy Baumeister point out that there is an analogy between superhero comic books and another sort of unrealistic simulation: "Much like the appeal of the exaggerated, caricatured sexuality found in pornography, superhero comics offer the appeal of an exaggerated and caricatured morality that satisfies the natural human inclination toward moralization."

But there are sophisticated moral tales as well. These can inform and entertain us with deeper insights about morality, capturing, say, the point that nobody really sees themselves as a villain. Other stories illustrate how good and evil can coexist in a single person, how bad results can come from good intentions, how revenge might not be sweet after all, and so on. To take one other example, the Coen brothers' films often show us that violence, even violence in service of a moral cause, cannot be contained—in movies such as *Blood Simple* and *Fargo* and *No Country for Old Men*, we see careful plans blow up in a terrible (and sometimes comic) manner.

At the same time, though, even the most highbrow fictions offer us the more primitive satisfaction of comeuppance. We enjoy watching bad people get theirs. You probably don't need the neuroscience here, but just for the sake of completeness, I'll note that areas of the brain associated with pleasure and reward are active when we witness individuals receiving fair treatment, but they are active as well when we see bad actors getting punished, even if their bad acts weren't directed at ourselves or anyone we care about.

This pleasure is grounded in sound evolutionary logic: if we weren't predisposed to punish or exclude bad actors, there would be no cost to being the turd in the punch bowl, and cooperative societies couldn't get off the ground. This logic applies with even greater force when it comes to an appetite for retaliation and revenge—if you don't deter others from preying on you and those you love, you are a perfect target for the unscrupulous and the psychopathic.

And so, it's hardly surprising that vengeance is a major

appeal in narrative, from classic works such as *Hamlet* and *The Iliad* to schlocky films like *An Eye for an Eye*, *Death Wish*, and *I Spit on Your Grave* to television series like the aptly named *Revenge*. I have two books on my shelf that purport to summarize most of the literary tradition; they are called *Revenge Tragedy* and *Comeuppance*.

I've been focusing on the pleasure of condemnation here, but, as a milder force, there is the pleasure of goodness. There can be praise and exultation and even awe for the hero, as well as the vicarious pleasure of imagining oneself in the hero's role. It's interesting, though, that this seems to be a milder pleasure than comeuppance, perhaps because there isn't the same evolutionary need for us to scrutinize and praise and take delight in goodness as there is for us to focus on the bad. As usual in psychology, the negative is more powerful than the positive. This is why good-versus-evil clashes are so much more satisfying than fictions where there is good without evil. I don't think the Batman movies would be quite so popular if, instead of fighting crime, billionaire Bruce Wayne devoted his resources, at great personal sacrifice, to building better housing and infrastructure in Gotham City. Heroes are fine, but we need the villains.

Once again, the pleasures of fiction—in this case, the desire to see justice done—are also the pleasures of reality. As I write this, a video of a New York lawyer named Aaron Schlossberg just came out. Schlossberg was recorded directing a racist rant at Spanish-speaking employees at a Manhattan deli, demanding they speak English and threatening to call the authorities on them and have them deported. People were outraged. Mobs of reporters followed Schlossberg on the street, screaming questions; a live mariachi band set up in front of his apartment, and he was kicked out of his office space. His pleading apology was derided.

I know many people who defend this, who argue that humiliation is necessary to deter ugly racist behavior, so, perhaps reluctantly, we carry it out. But if you look at tweets and Facebook posts, or at the faces of those protesting Schlossberg on the streets of New York—many of whom are progressives, the sorts of people who explicitly disdain vengeful impulses—you'll see *glee*. People enjoy watching Schlossberg get what he deserves.

This is similar to what happened when Richard Spencer, the famed white supremacist and founder of the alt-right movement, was punched in the head while being interviewed in Washington, D.C. For many, this was the most perfect thing ever; some of the tweets I saw took the clip of Spencer being assaulted and set it to music.

You might detect a bit of disapproval on my part, and, yes, I'm against public shaming and physical assault, even for the very worst people. One reason for my hesitancy here is that, even putting aside the broader moral issues, it's remarkable how often the shamers and assaulters end up attacking the wrong person or getting the facts entirely confused. (For many examples of this, check out Jon Ronson's book *So You've Been Publicly Shamed*.) And even when they are right, they can be wrong; as then-Yale graduate student Matthew Jordan and I have discussed, group responses to individual transgressions, especially over social media, are often grossly disproportionate. It's often enjoyable to mock someone online, and it feels like a minor act, but when you multiply this by thousands, the effect can be terrible.

I'm not here to nag, though; my point is just to remind you

of how much pleasure we take in real-world comeuppance. I have a friend, an evolutionary psychologist, who likes to ask people if they've ever wished that someone they know personally would die. I've started to ask this of people myself, and I've gotten a lot of yeses, and even more if I ask about making someone suffer. Often, they want the person to suffer and to know why they are suffering, to have justice done and to see justice done. In *The Princess Bride*, it's not enough for Inigo Montoya to kill the man who murdered his father; he also has to give him a little speech first. "Hello, my name is Inigo Montoya. You killed my father. Prepare to die."

People differ in the extent of their retaliatory drive, with regard to what they like to see in both fiction and the real world. There are many people who spend a lot of time wanting to hurt those they think have hurt them or a loved one, often for what seem to others to be fairly minor infractions. Some are less bloodthirsty but still feel this desire. A friend of mine once told me that she never wished anyone dead or wished anyone to suffer, not really, but she later confessed to me that she got some pleasure out of imagining certain individuals with some difficult affliction—in particular, just a dash of urinary incontinence.

This third explanation for the pleasure of aversive fiction, then, is that we like good acts and we particularly like comeuppance, and so the suffering caused by a villain is, well, a necessary evil. It sets the stage for the pleasurable payback.

But there's another sort of appeal as well. We find evil fascinating. Everybody knows that the most interesting character in *Paradise Lost* is Satan—it's often said that he gets all the best lines—and who would doubt that the Joker is more compelling than Batman, that Hannibal Lecter is much more charismatic than Clarice Starling? Even our heroes these days tend to be antiheroes—criminals, rogues, people with a dark past.

The attraction of such characters has many explanations, but I think that a major factor here is that some of us occasionally do wish to dominate and control, to be feared, to get what we want. We might enjoy enacting these fantasies in imagination. Who isn't occasionally jealous of the psychopath, so unencumbered by guilt and shame and worry? Some genres help this along by making the bad guy more appealing, often charming and interesting. But even without that, there is a draw to evil, and this is yet another appeal of fictions in which bad things happen and good people suffer.

4

STRUGGLE

How much do you have to pay people to engage in embarrassing, painful, or immoral acts?

This question was the focus of one of my favorite scientific articles, written by the psychologist Edward Thorndike in the 1930s, called "Valuations of Certain Pains, Deprivations, and Frustrations." Psychologists tend to identify this period with technical work done by stern older men with unfortunate facial hair, but one look at the paper and you know that it's something special.

The study was simple. Thorndike made up a long list of unpleasant activities and asked his subjects—students and teachers in psychology and young unemployed people—how much money they would need to do them.

He doesn't tell us why he is doing this. He presents no specific hypothesis and no practical problem he hopes to address. He just thinks it's an interesting question and assumes that we'll agree. He begins by saying he wants to learn how people think about *disutilities*—"pains, discomforts, deprivations, degradations, frustrations, restrictions, and other undesired conditions"—and says that this topic is "obviously important." And then he's off to the races.

Thorndike was an old-school empirical scientist, and he concedes right away that the best way to explore this issue would be through experiment and observation. But then he adds that asking people about money is "by no mean valueless, if used reasonably." He couldn't have known that asking people to put prices on things would become common in social psychology, behavioral economics, and other areas of psychology. After all, it's one thing to learn that people think X is bad, and another to learn that people think Y is bad, but often you want to compare the two. Getting your subjects to put dollar values on X and Y—how much they would pay to get them; how much they would pay to avoid them—is a good way to find this out.

Thorndike asked about hypothetical payments; no money changed hands. Some behavioral economists would insist, with some justification, that this is an imperfect method—what people say they're willing to pay and what people actually pay are quite different. But to do this with real money would be too expensive, and anyway, as you're about to see, it couldn't be done for the sorts of scenarios that Thorndike was doing. It's psychology, not torture porn.

His scenarios involved physical pain and mutilation ("have one upper front tooth pulled out"), harm to others ("choke a stray cat to death"), constraints on the rest of one's life ("have to live all the rest of your life in Russia"), and constraints *after* the rest of one's life ("lose all hope in life after death"). There were disgusting acts ("eating a quarter pound of uncooked human flesh"), taboo acts ("spit on a picture of your mother"), and embarrassing acts—some of which illustrate how much times have changed ("walk down Broadway from 120th Street to 80th Street at noon wearing evening clothes and no hat").

In an earlier book, I used a finding from this survey to illustrate an important point about our psychologies. When Thorndike asked how much money people needed to strangle a cat, the average answer was $10,000 (or about $185,000 in today's dollars). This is a lot—over twice as much as they required in order to have one of their own front teeth pulled out. People would rather be painfully disfigured than have to kill a harmless animal with their hands! This contrast between self-harm and harm to others has recently been expanded upon by my colleague Molly Crockett, who has used a variation of Thorndike's method to show that, under certain circumstances, people would prefer to be shocked than to shock an innocent person. If someone tells you that people care only about their own well-being, such findings make for the start of an excellent rebuttal.

We can use the Thorndike method to explore the focus of this chapter—work and effort. Just to start off, how much would you need to be paid to do each of the following?

Spend one hour moving furniture from an apartment into a moving van.

Spend two hours moving furniture from an apartment into a moving van.

It's not cat strangling or tooth extraction, but you likely would want more money for the second than for the first. (This is certainly the case for professional movers.) This might reflect the value of your time, but it isn't all that matters. Consider this contrast:

> Slowly move light furniture into the van for an hour.
> Slowly move heavy furniture into the van for an hour.

These two jobs take the same amount of time, but the second is harder, and people would presumably ask for more.

Well, duh. When you bring your car into the shop, you're charged for "parts and labor," and you've never questioned for a second that the more labor it takes, the more you have to pay. Indeed, the relationship between effort and financial cost is so tight that we often talk about our everyday efforts in economic terms—we invest effort, labor over a decision, pay attention, find certain activities taxing, and so on.

These simple examples show that *effort*—technically defined as intensification of either mental or physical activity in the service of meeting some goal—is something you have to pay people to do. And this is because people typically don't want to engage in effort; it is, to use Thorndike's stilted phrase, a *disutility*, falling into the same category as embarrassment, pain, and morally forbidden acts.

The difficulty and unpleasantness of effort increase as time goes on. The second hour at a task is typically harder than the first hour, and the third hour is harder than that, until it's too much and we have to stop. We can work on effortful tasks for only so long before we just run out of gas. This increase in difficulty doesn't reduce to general physical exhaustion. You might get really tired of one task, like completing tax forms or putting together Ikea furniture, but have plenty of energy to go for dinner with friends or wrestle with your toddler. You're not tired in general; you're tired of expending a particular type of effort.

Quick interlude: I know all of this so far is obvious. But that's okay. Much of this book is about the odd and the unexpected, but the obvious needs to be explained as well. Physicists don't just explain black holes and quantum anomalies, they also need to explain why apples fall and water turns to ice when it gets cold, phenomena that everyone is familiar with. So, too, for psychologists. In 1890, William James had some thoughtful things to say about this:

> Why do men always lie down, when they can, on soft beds rather than on hard floors? Why do they sit round the stove on a cold day? Why, in a room, do they place themselves, ninety-nine times out of a hundred, with their faces towards its middle rather than to the wall? Why do they prefer saddle of mutton and champagne to hard-tack and ditch-water?

Most of us don't normally reflect on such things, he says. But some do:

> It takes, in short, what Berkeley calls a mind debauched by learning to carry the process of making the natural seem strange, so far as to ask for the why of any instinctive human act. To the metaphysician alone can such questions occur as: Why do we smile, when pleased, and not scowl? Why are we unable to talk to a crowd as we talk to a single friend?

Right now, let's be the metaphysicians debauched by learning. We will munch on some saddle of mutton and sip some champagne and try to make sense of the mundane and the obvious before returning to the strange.

EFFORT TAXES THE body and the soul. People doing effortful tasks talk about anxiety, stress, and frustration—all bad things. If you give subjects in a lab an effortful task, it usually leads to higher blood pressure, sweating, and pupil dilation, reactions associated with activities we don't enjoy. Effort is also associated with a certain facial expression involving contraction in the corrugator supercilii muscles near the eye—in other words, as you work, your face scrunches into an unhappy expression. And the extent of one's effort is related to activity in the anterior cingulate cortex that typically corresponds to aversive and unpleasant activities.

Nonhuman animals, who can't tell us about what they feel and can't answer the Thorndike questions and don't have faces that scrunch up in the right way, also dislike effort. If you set up a maze so that there are two ways to get to the food, one easy and one hard, rats will choose the easiest one. If a foraging animal can easily get food in area A and needs to struggle and work for food in area B, which area do you think they'll hang out in?

Based on the animal research, psychologists proposed long ago a psychological law about effort. The law of least work says that, given a choice between similarly rewarding options, organisms, including us, avoid those that require more effort.

The cost of effort shows itself in all sorts of ways. Suppose you need to signal your love to someone. Evolutionary psychologists, animal behaviorists, and advice columnists know full well the value of the costly signal. Such a signal is hard to fake; it cannot be sent unless you both have the resources and are willing to bear the cost. (We talked about costly signals in an earlier chapter when discussing self-harm.) Critics who complain that engagement rings are expensive and useless and that the money would be better spent on practicalities like saving to buy a house are missing the point. Expensive and useless are what engagement rings are for, because this means that it hurts to buy one. Nobody has ever said "I love you so much that I'll eat a hot fudge sundae for you," because many of us would eat a hot fudge sundae even if we *didn't* love that person. It's the opposite of a costly signal.

The costly signals of human courtship often involve money, but they're ultimately about sacrifice, so sustained effort will certainly do the trick. Consider the song by the Proclaimers with these wonderful lyrics about the extent of the singer's love:

But I would walk five hundred miles
And I would walk five hundred more
Just to be the man who walks a thousand miles
To fall down at your door.

Effort and then more effort on top of it! So romantic. The gifts I remember the most are the ones that took the most time and effort and sacrifice on the part of the giver, either to acquire them or to create them. They expressed their commitment and affection to me.

(I can't resist a qualification about the five-hundred-miles claim, though. Talk is cheap; the singer is asserting, not doing. Maybe that's why they call the group the Proclaimers?)

We tacitly appreciate the law of least work when making sense of others' choices. I've been enjoying Japanese whiskey these days—and you'd know about my preference if you watched me at Frank's Liquor Store, around the corner from my house. But if Frank's has run out and getting my Suntory would mean driving across town, I'll buy Macallan instead. And you would infer from all this, correctly, that I like Japanese whiskey more than Scotch, but not so much more that it's worth the extra work to get it. Effort has a cost, and this cost factors into how we make sense of what people do.

THE EXAMPLES OF effort so far have been physical: schlepping things around and walking and driving across town. But mental activity can be difficult as well, and when we talk about effort, we don't usually make a distinction between body and mind.

Here's what mental effort feels like. Please remember the number 7. That's easy. Now remember my phone number from when I was a teenager: 555-688-9058.* Keep this in mind for the next five minutes. Doing this is difficult and maybe a bit annoying, the psychological equivalent of holding a small dumbbell over your head. To take another example, choosing from among fifteen flavors of ice cream is harder than choosing from three. Indeed, there is a whole literature on the "paradox of choice" that focuses on the stress associated with difficult decisions.

One psychology experiment, done more than a hundred years ago, explored the extremes of mental effort. As part of her

* I changed the area code; my father still has the same phone number.

doctoral dissertation, called *Mental Fatigue* (done under the supervision of . . . Edward Thorndike!), Tsuruko Arai performed a series of grueling experiments on herself, involving multiplying pairs of four-digit numbers in her head. She spent four days, twelve hours per day, solving these problems constantly. Arai found that this got more difficult as time went by, and she concluded that "difficult and disagreeable continued work brings about a decrease in the efficiency of the function exercised." That is, psychological effort shows just the same pattern as physical effort.

Decades later, another team of researchers replicated Arai's findings, with the ingenious methodological twist that they got their graduate students to do the unpleasant ordeal instead. Three students went through her nightmare routine. They did worse as the days progressed, though the decline was not as dramatic as in Arai's original study. What was really clear was that the students hated it, reporting exhaustion, restlessness, and boredom. One said, using a phrase that would have pleased Thorndike, that she "would not repeat these four days for $10,000."

The exercise of mental effort is intimately related to self-control or willpower—the deliberate overriding of some other, more tempting response. In a sense, every effortful mental task, like every effortful physical task, is a willpower task, because every task involves overriding the desire to do nothing, to overcome inertia.

Willpower is of obvious importance in everyday life. If I had the power to raise everyone's intelligence, I would do so in a second, because intelligence is related to all sorts of good things, such as better planning for the future and increased kindness to others. But if I had the option of a magic spell that could increase self-control, that could make mental effort come easier, I would be even more enthusiastic. Deficits in self-control (also called failures of impulse control) lead to addiction, crime, relationship problems, and so much more.

Also, self-control just makes life better. Who doesn't wish they had more willpower when faced with unhealthy food and other temptations? Wouldn't it be nice to never lose your temper and never get distracted by social media? Wouldn't it be amazing to be able to work on a project for however long you wanted to, to be unmoved by the distractions of email, the refrigerator, and the sofa?

Some of us look for ways to deal with failures of willpower. Just as a random crazy-ass example, take writing a book. My schedule for writing this book is an hour a day, first thing in the morning. I do it early so I get my hour in before other distractions and obligations arise. Also, I'm a morning person, workwise: I can make much more use of a free hour at eight in the morning than at two in the afternoon. (Surveys of productive people suggest that this is a typical pattern: many people do their best work in the morning, somewhat fewer do well late at night, and fewer still are at their best in the afternoon.) Sometimes I can work for longer, but on some mornings I can't even do my hour, so I shift to different strategies of alternating work and not-work, sometimes using a modified version of the Pomodoro Technique, where I work for a few minutes on one thing, a few minutes on another, bit of book, bit of email, bit of Twitter, bit of class preparation, embedding my one hour of writing in three or four hours of activity. (I know this isn't for everyone. Everyone has a superpower; mine is that I can work in eight-minute bursts.)

I didn't always know that I am most productive at difficult

tasks in the early morning, that for me this is the best time for what Cal Newport calls "deep work." It was a useful thing to learn. But still, I'm frustrated by my limitations. I sometimes have a whole day free, and if I could work continuously throughout that day, or even for most of it, I could get this book done before my deadline. Think how happy my editor would be! But I just can't.

And yet, this isn't quite right; it's not *literally* that I can't. It's a similar situation to being physically exhausted. If I tell you that I'm so tired I can't walk another step, and you say, convincingly, that you'll give me a million dollars to walk a mile (or point a gun at me and say you'll shoot me if I won't), then I'll walk that mile. What's closer to the truth is that further work is *hard*—effortful and unpleasant.

THE LAW OF least work makes sense for physical work because bodies can be damaged by overuse. It's not that the physical constraints of the body actually make one stop working. Muscles tire, backs ache, feet get sore, but, with the possible exception of weight lifters "training to failure," nobody stops working because they literally cannot do more. Rather, the experience of physical exhaustion reflects, at least in part, the output of a system that monitors stress to the body—if you overexert yourself, you can cause damage, so it makes sense to have a system in your head to say "Slow down" and then "Stop" when the strain is too much.

But why would mental effort be similarly difficult? If you think too hard, you're not going to pull a muscle or snap a bone; there is nothing like that in the brain. So why can't we keep at mental work for as long as we want?

One possible explanation has been advanced by Roy Baumeister and his colleagues. They posit that mental effort (or self-control, willpower, or grit) actually is a lot like a muscle. Like a muscle, it can work for only so long before it gets tired; like a muscle, it can be strengthened through exercise.

This view nicely captures the fact that, just as we differ in our actual musculature, we also differ in strength of willpower. There are intellectual mesomorphs who have seemingly limitless powers of concentration, and there are ninety-eight-pound cognitive weaklings who can't focus for even a minute. Being high or low in willpower seems to be a general personality trait, and, as I mentioned before, being low in willpower is trouble—it makes one more likely to smoke, get into car crashes, have unwanted pregnancies, and commit crimes.

But, again, since there aren't muscles in the brain, why do we get mentally tired? Perhaps, like muscles, brains feed off a limited resource. Baumeister and his colleagues suggest that this is glucose (sugar). This theory is supported by the fact that sugar does seem to have an energizing effect. Running out of steam? Have a candy bar.

This limited-resource theory of willpower has been influential. It has spawned best-selling books, including one by Baumeister and John Tierney called *Willpower*. One piece of advice they offer is that one should be careful not to use up one's willpower on unnecessary tasks. You wouldn't tire out a muscle before a weight-lifting competition, would you? I knew this research was having an impact when I saw that Barack Obama, when he was president, was paying attention to this particular piece of advice. He talked about it during an interview with Michael Lewis:

> "You'll see I wear only gray or blue suits," he said. "I'm trying to pare down decisions. I don't want to make decisions about what I'm eating or wearing. Because I have too many other decisions to make." He mentioned research that shows the simple act of making decisions degrades one's ability to make further decisions. "You need to focus your decision-making energy. You need to routinize yourself. You can't be going through the day distracted by trivia."

(This interview took place shortly after Anthony Weiner had tweeted, by mistake, lewd pictures of himself, so the magazine *Reason* summarized Obama's musings about self-control with the headline OBAMA WEARS BORING SUITS SO HE WON'T TWEET PICTURES OF HIS PENIS.)

THIS BRAIN-AS-MUSCLE THEORY captures everyday experience in an elegant way. But it has some serious limitations.

The claim about glucose is the weakest part here. It is unlikely that the exercise of intellectual effort reduces glucose levels in the brain in any nontrivial way. Hard mental tasks don't actually burn more glucose than easy mental tasks. Indeed, what really drops glucose is exercise—but, contrary to the predictions of the glucose hypothesis, exercise tends to make you better at subsequent tasks requiring mental effort, not worse. Critics have also pointed out that the biggest drops in glucose consumption corresponding to brain activity aren't from cognitive load; they're from opening one's eyes. But we don't perceive this as difficult or effortful.

Now, most psychologists would agree that getting glucose into the system—eating or drinking sugar—tends to up your performance on effortful tasks. But it's long been known that glucose influences reward circuitry in the brain. (Sugar is a drug, man!) And this would be how it's having its effect, not through any calories it brings into the system.

A more promising alternative to the limited-resource account has been developed by Robert Kurzban and his colleagues. This involves a notion from economics, that of "opportunity cost," standardly defined as "the loss of potential gain from other alternatives when one alternative is chosen."

Suppose I agree to review a textbook for $300, but it takes eight hours to do so. Smart move? Well, in part it depends on what I think about reviewing textbooks and how much I like money, but it also depends on what else I could be doing in that time. If it turns out that I could make twice as much doing something equally unpleasant, then, other things being equal, I should say no. In general, if another activity exceeds the benefit of your current activity, then you should stop your current activity and do that other thing.

This is a theory of why effort is often unpleasant. The phenomenology of getting tired doesn't reflect a diminishing resource; rather, it is about growing opportunity cost. This feeling of difficulty is a signal that there are better things to do elsewhere.

This theory has the promise of explaining why only some activities wear us out. Looking out the window doesn't feel effortful because this mental activity doesn't soak up capacities that you could be using for other things; there are no opportunity costs. Listening to classical music doesn't exhaust me because I can do it while I check my email. Compare this with moving boxes or adding numbers in your head. *These* are tiring, because they take you away from other activities, and so they gnaw away at you. The fatigue of effort is a neural reflection of FOMO—fear of missing out.

From this perspective, the cost of effort isn't a glitch in the system. It's valuable, something we would want to include if we were building a humanoid robot that could survive on its own. If someone could effortlessly spend unlimited hours on any task they were engaged with, it would cause them to miss other things, such as social stimulation and social contact. This speaks to why effort gets harder over time—it's not that a finite pool is being used up; it's that, as the hours go by, the value of other activities grows.

To provide a somewhat strange analogy, consider the sexual refractory period. While one's mileage may vary, depending on age and sex and occasion, returning to the arousal-and-orgasm process typically requires a wait, and soon you're not in the mood anymore. Maybe the body is just built that way; maybe it's a glitch in the system. But it is also quite adaptive. Orgasm can feel so good that, without this mechanism, some would never stop pursuing it, giving up all other worldly pursuits. Evolution, being smarter than we are, compels us to take a break and do other things.

WE TALKED ABOUT why effort is hard and unpleasant. But this is a book about the appeal of suffering, and now, armed with what we've learned so far, we can get to what Michael Inzlicht and his colleagues have called "the effort paradox."

They start with what we've discussed so far; there is abundant evidence supporting the law of least work, showing that humans and other animals don't like to work, don't like to apply effort, don't like to exercise willpower. But then they point out that sometimes the opposite occurs. We will often choose to do something rather than nothing, even if the something is effortful and provides no tangible benefits. Effort itself can be a source of pleasure.

It has long been known that effort can be the secret sauce that makes things better. One of the classic findings in psychology is that the more effort you put into something, the more you value it. This is the logic of Benjamin Franklin's classic advice on how to turn a rival into a friend—ask him or her to do you a favor. Having worked to help you, they'll like you more.

Or remember Mark Twain's story of when Tom Sawyer had to whitewash his aunt's fence. When Tom's friends come by, he pretends to be delighted at this task, and soon his friends end up paying him for the privilege of working on the fence—and they seem to really enjoy themselves. As Twain puts it, Tom Sawyer "had discovered a great law of human action, without knowing it—namely, that in order to make a man or a boy covet a thing, it is only necessary to make the thing difficult to attain."

Effort sweetens the value of the products of labor. When instant cake mixes were introduced in the 1950s, housewives initially rejected them for being too easy; the manufacturers changed the recipe so that you had to add an egg, and then they became more appealing. As I write this, meal delivery services are popular; these services send you a small assortment of ingredients and simple instructions and then you make your own dinner with them. It might well be that this is healthier and less expensive than being sent ready-made food, but my hunch, based on the fact that many who use these services are wealthy enough that they can order fully prepared food, is that what they offer is the satisfaction of cooking your own meal. Similarly, living in Connecticut, I've done my share of apple picking, peach picking, and berry picking, and I can assure you that food you pick yourself really does taste better—and it's not just because it's fresher.

So far this is all anecdotal, but Mike Norton, Daniel Mochon, and Dan Ariely did a series of studies where they asked people to build things—such as folding origami and putting together Lego constructions. They found that people will pay more (in one study, *five times* as much) for what they themselves have created than for the same creation made by a stranger. This effect exists even when there's just one way to complete the task, so there's no room for creativity or for a personal touch, and even when doing the assembly is actively unpleasant. In a shout-out to the Swedish big-box store, they call this "the Ikea effect."

One common explanation for this association between effort and value is that our minds are sense-making machines—we seek to justify our own actions. If I ran naked through the quad to get into the fraternity, it must be a damn good fraternity. If I went to the effort to make these stupid origami things, they must be pretty special.

But the sense-making theory—also known as the cognitive dissonance theory—cannot be entirely right. One concern is that it predicts that only our own effort should sweeten our activities. But the same phenomenon holds for others' efforts as well. In one series of studies, people were shown a poem, a painting, and a suit of armor, and were told that it took varying amounts of time to create these things. As was predicted, those who were told that it took more time to create an item liked it more and thought it was better. This runs contrary to the sense-making theory.

Also, nonhumans show the same effect. Rats, for instance, will press a lever for longer to get food that they had previously worked to obtain than to get food that came easy, suggesting that they value the harder-to-get food more. Similar findings come from a range of creatures, including, in one recent study, ants! Now, I'm willing to accept that ants are plenty smart, but I don't think they're going through some complicated psychological process to make themselves feel better about past choices.

As Inzlicht and his colleagues propose, the animal findings suggest that effort can become pleasurable through simple association. Suppose whenever your dog does something you want to reward, you take out a treat and, right before you give it over, you say "Good dog!" Before you start practicing this routine, the dog enjoys the treat—this pleasure is unlearned—and "Good dog!" means nothing. But, as any animal trainer or Psych 101 student will tell you, the phrase becomes associated with the treat, and soon you can get the dog to shudder with delight by simply saying the words.

Now consider our own lives. The way the world works is that rewarding experiences often require effort. By the same logic of association that we saw with the dog, the effort (which might begin as negative) gets paired with the reward and then becomes rewarding in itself. If you suffer for something that gives delight, soon the suffering itself can give joy.

At least for humans, though, this is not the whole story. Jeremy Bentham talked about the pleasures of mastery, and I think, from having observed small children, that this is something we start off with. We enjoy certain sorts of effort for their own sake. I would speculate, though I admit I don't have evidence for it, that certain kinds of effort *fundamentally* tickle us, that we are wired so that the right degree of struggle, regardless of the results, gives us deep satisfaction.

LET'S DIVE DEEPER into the sorts of effort that people enjoy. To take a perfectly mundane example, I like doing the *New York Times* crossword puzzle. It is intrinsically pleasurable. Nobody pays me to do it, and I'm not good enough to ever impress someone with my skills. Yet it's fun.

There's some pleasure in finishing the puzzle, but this isn't the main draw. I can easily complete the Monday puzzles, but I don't bother with them; they're too easy. It's the later days of the week, when they get harder and when I often fail, that I enjoy. Successful endings aren't as satisfying or essential as many think. Remember: Rocky lost.

The struggle is at the core of the pleasure here. Think about crumpling paper and tossing it, from a distance, into a wastebasket, trying to get three in a row. Or eating M&Ms as a couple, one throwing the candy through the air, the other trying to catch it in his or her mouth. Such activities have no inherent value; we invent them to make mundane activities such as disposing of garbage and eating sweets interestingly difficult.

What makes some activities exceptions to the law of least work? We talked before about explaining the ebb and flow of effort in terms of opportunity cost; we feel tired and bored when there is something better to do; that's why effort is typically aversive. And so we can frame the question like this: What is it about certain types of effort that make them better than possible alternatives? What distinguishes furniture moving from crossword puzzles?

One answer is that effort becomes enjoyable when it's seen as play, or as a game. ("I don't think of it as *work*," the productive person says in an interview about the secrets of her success.) There is a whole movement of "gamification," where something that is plainly not a game is presented as one.

This doesn't answer the question; it just reframes it. Now, instead of asking what sort of effort is pleasurable, we're asking what sort of effort counts as a game? Still, the reframing is useful, because people have thought deeply about what makes for a good game. The following properties often come up.

1. **An attainable goal**. There might be a pleasure in the exercise of what philosophers call *atelic* activities, things that you can never complete because they have no end. ("Telos" means purpose; "atelic" means without purpose.) Think of an aimless walk or just goofing around with friends. But in general, it's hard to sustain effort at atelic tasks, particularly if they are difficult or aversive. In Greek mythology, Sisyphus had to roll an immense boulder up a hill, then have it roll down to the bottom, then roll it back up again, forever, and it's easy to see why this is such an awful fate—there's no goal; he's never finished.

   Having a goal might be a necessary condition, but it's far from a sufficient one. Many effortful tasks that nobody would do for fun, like cleaning a bathroom, have tangible goals. But still, the goal does add to the pleasure of the experience. Cleaning the bathroom for thirty minutes is unpleasant, but wouldn't it be worse to spend a half hour cleaning a Sisyphean bathroom, one that stayed dirty no matter how much you scrubbed it?

2. **Sub-goals, some indication of progress.** Part of the pleasure of a crossword puzzle is the feeling of progress as you get closer to completion, bit by bit, through the meeting of small goals. This is what much of gamification is about: using points or currency or badges or progress bars to indicate that you're getting closer to the end; these provide little jolts of reinforcement. When I run, I use a GPS watch and it turns the run into a sort of game. I can compare my pace and time across multiple runs, trying to make progress, but, more than that, I can establish sub-goals through the run—*Okay, I'm going to pick up the pace to X for one minute, then reward myself by dropping down to Y for the next minute*—and it does make the time go faster and the experience more pleasant.

   This brings us closer to the idea of a pleasurable activity, but we're still not fully there. Moving furniture is full of sub-goals—the hundred pieces of furniture you have to move into the van are a hundred sub-goals—but it's still not pleasurable.

3. **Mastery.** The right game establishes an optimal level of difficulty. Many video games, including simple ones like *Tetris* and *Angry Birds*, start off easy and get increasingly challenging, so that you end up spending most of your time at a point where there's just the right amount of struggle. There is a pleasure in getting good at something, at doing better than you once did and better than other people do. This distinguishes enjoyable activities from those such as cleaning bathrooms and moving furniture, activities that, for most people, at least, don't fall into this optimal level of effort.

4. **Social contact, camaraderie, and competition**. None of these are essential. Many perfectly good games are solitary. But still, playing with others and against others does add a lot. The most popular video games are those that are played in teams and in competition. This is often one aspect of gamification: when work is transformed into a game, people are sometimes given avatars and can see the avatars of others; they can form teams; and their scores can be marked on leaderboards.
5. **Collections**. Some popular and enjoyable games involve acquisition, as when you try to collect all the Pokémon in *Pokémon Go*. This can be seen as just a specific form of "sub-goals" (property 2 above), in that each item you collect reflects the satisfaction of a sub-goal, but perhaps there is some distinct satisfaction going on here in acquisition that you don't find in crossword puzzles or first-person shooters.

These are all features of enjoyable games, but we'll see some of them reappear in the next chapter, where we move away from pleasurable pursuits and talk about meaningful ones.

THE BEST ANALYSIS of what sort of effort appeals to us was developed long before anyone ever used the word "gamification." This comes from Mihaly Csikszentmihalyi's insights about the nature of *flow*. For Csikszentmihalyi, flow is an experience of intense and focused concentration, where you are entirely in the moment.

I first encountered his work when reading *Flow: The Psychology of Optimal Experience*. Much of this book comprises descriptions of what flow is like, based on his interviews with chess players, dancers, rock climbers, and others. As I mentioned earlier, this book had a big influence on me, helping me make sense of what I found satisfying in my life and what I didn't. (It revealed to me what I knew but didn't know that I knew.) It also made me envious of the people Csikszentmihalyi interviewed, who were lucky enough and gifted enough to spend so much of their lives in flow states.

How does one get into such a state? These are Goldilocks experiences: not too cold, not too hot. There is a sweet spot here: you have to be challenged to just the right extent, threading the needle between too easy (which leads to boredom) and too difficult (which causes stress and anxiety). Flow experiences typically involve clear goals with immediate feedback about your experiences. And so the first three criteria I listed immediately above—goal, sub-goals, and mastery—fit nicely into this analysis.

Typical examples are complex athletic activities (rock climbing) and intellectual activities (writing). But while there are experiences that usually facilitate flow and those that usually impede it, a lot of it depends on the individual. You can imagine two people in a discussion, and one is in an exquisite state of flow, challenged to an optimal degree, rapt and lost in the moment, while the other is panicked and overwhelmed—or bored out of her skull.

There are dramatic differences in how much flow people have in their lives. Some will have none even if they are in a world that has abundant opportunities. Others will work to create it even in the worst of situations, such as solitary confinement. Csikszentmihalyi talks about people with "the autotelic personality," who do things for their own sake and aren't chasing external goals. Such a personality is associated with traits like curiosity, perseverance, and what Csikszentmihalyi calls "low self-centeredness," a state in which you are less focused on yourself and how you are seen by others. (Not thinking about what others think presumably allows for better immersion and focus.)

Few are fortunate enough to have lives of flow. Along with Jeanne Nakamura, Csikszentmihalyi did surveys on Americans and Germans and found that about 20 percent often experienced flow states, which they defined as involvement so intense that they lose track of time. But many more—over 33 percent—said that they never experience such states. Other studies find that although some people get little out of non-flow, low-challenge activities (doing the easy crossword puzzle, watching trash television), others like them just fine and avoid the alternatives.

If there is such satisfaction in flow, why do some people so rarely enter that state? One problem is that it is hard to get started. It's easy to do easy things, after all. Easier to sit on the sofa than to put on your running gear; easier to watch YouTube than start a challenging intellectual project. And, once you're in flow, it might be hard to stay. The sweet spot might be elusive. To put it in the framework of the opportunity cost theory described above, other activities begin to attract you with increasing intensity.

Also, as time goes on, there are diminishing returns on certain tasks. In writing, good ideas might initially come to you, but soon the low-hanging fruit is picked and putting new words on the page becomes harder and more frustrating. Working on a crossword puzzle gives you the satisfying thrum of working to find the right answers, but as the level of difficulty increases, you might move past the flow state into frustration.

Flow is wonderful, then, but it's difficult to find—sandwiched between boredom and anxiety, hard to get started, hard to sustain.

CSIKSZENTMIHALYI EMPHASIZED EMPLOYMENT as a major source of flow. But even if this were true when he wrote the book, more than thirty years ago, it might not be true now. A poll from Gallup asked more than 200,000 people from 142 countries about their work. They were instructed to sort themselves into one of the following three categories:

1. **Engaged employees:** "Engaged employees work with passion and feel a profound connection to their company. They drive innovation and move the organization forward."
2. **Disengaged employees**: "Not engaged employees are essentially 'checked out.' They're sleepwalking through their workday, putting time—but not energy or passion—into their work."
3. **Actively disengaged employees:** "Actively disengaged employees aren't just unhappy at work; they're busy acting out their unhappiness. Every day, these workers undermine what their engaged coworkers accomplish."

Only 13 percent said they were engaged, 63 percent said that they were disengaged, and 24 percent saw themselves as actively disengaged. Put simply, many people think their jobs suck.

There are all sorts of reasons why this is so. Many jobs have degrading conditions, perceived unfairness, and lack of autonomy. In the language of David Graeber, many people spend much of their waking life doing bullshit work on bullshit jobs—work that feels pointless, unnecessary, or pernicious. At the moment-to-moment level, there is little engagement or flow; at the broader level, there is a lack of meaning and purpose.

Some jobs, though, are relatively bullshit-free. Some are associated with meaning. In one survey, more than two million people were asked what they do and then asked about how much meaning they have in their life.

It turns out the most meaningful job is being a member of the clergy. This is followed by serving in the military, being a social worker, and working in a library. This is an intriguing list. All of these jobs involve a lot of personal engagement and some amount of difficulty. The pay isn't great, and they are not very high status. If you want both meaning and money, your best bet is to become a surgeon, which pays very well, is high status, and is seen as very meaningful. But that's about it.

(The least meaningful jobs? Food preparation and service and sales rank very low. The job that has the lowest perceived meaning of all? Parking lot attendant.)

It's not just the job, though. Two individuals can take the same job and react to it in different ways. I see my job as a professor as having a lot of freedom, a great environment, and productive and engaging day-to-day work; I can't imagine a more fulfilling calling. But in the last few years, three people I know retired early from tenured jobs at top universities. Unlike me, they found the work boring and unsatisfying and frustrating.

On the flip side, Csikszentmihalyi notes that just about any job can be transformed into a meaningful one. It's a trope of the Buddhist tradition that menial labor—scrubbing toilets, say—can, when done by the right person, take on significance and value. Amy Wrzesniewski and Jane Dutton interviewed janitors at a large hospital unit and found a great range in how satisfied they were with their jobs. This was connected to how much the janitors saw themselves as part of the healing process. The value of connecting one's work to a broader purpose is nicely summarized by a story told by Emily Esfahani Smith about President Kennedy touring NASA in 1962 and talking to a janitor. When he asked the man about his job, he said that he was "helping put a man on the moon."

As you can tell, I'm a big fan of flow. It is associated with psychological health, it is personally rewarding, and it is connected with capacities that are good for one to have, such as focus and discipline. But its importance can be overstated. Jeanne Nakamura and Csikszentmihalyi wrote, "What constitutes a good life? . . . Flow research has yielded one answer, providing an understanding of experiences during which individuals are fully involved in the present moment. Viewed through the experiential lens of flow, *a good life is one that is characterized by complete absorption in what one does*" (italics theirs).

But this is actually a poor answer to the question of what constitutes a good life. Flow can be trivial. Crossword puzzles can be, for me and others, flow experiences, but a life of doing crossword puzzles would be a huge waste. And actually, as Csikszentmihalyi himself has discussed elsewhere, flow can be a lot worse than trivial. Adolf Eichmann, a key organizer of the Holocaust, has often been described as someone with no malice toward the Jews. He was an expert who sought to do a good job, and he took pride in mastering a complex and technically challenging task. Csikszentmihalyi suggests that Eichmann "probably experienced flow as he shuffled the intricate schedules of trains, making certain that the scarce rolling stock was available where needed, and that the bodies were transported at the least expense. He never seemed to ask whether what he was asked to do was right or wrong. As long as he followed orders, his consciousness was in harmony."

And here we see the limits of flow. After all, what sort of life would it be without some sort of purpose? What sort of life would it be without goodness and meaning?

5

MEANING

The best origin story doesn't come from religion, myth, or science. It's from *The Matrix*, where Agent Smith tells Morpheus how the world they are experiencing—a simulation created by malevolent computers—came to be:

> Did you know that the first Matrix was designed to be a perfect human world? Where none suffered, where everyone would be happy. It was a disaster. No one would accept the program, entire crops were lost. Some believed that we lacked the programming language to describe your perfect world, but I believe that, as a species, human beings define their reality through misery and suffering. So the perfect world was a dream that your primitive cerebrum kept trying to wake up from.

"Define their reality through misery and suffering." This phrase nicely captures an enduring idea—the stuff of theology, philosophy, and a million dorm room debates. And it nicely captures a key theme of this book, which is that some degree of misery and suffering is essential to a rich and meaningful life.

Meaning is a difficult topic to write about. There is a famous remark by the physicist Wolfgang Pauli, dismissing the work of another scientist: "He isn't right. He isn't even wrong." I often think about this line while reading about meaning and purpose. The problem isn't usually that I disagree with what I read—it's that it's too fuzzy and vague and general to take seriously. So now, as we reach the part of the book that's going to discuss these issues, let's see if we can make some claims that, if not true, at least rise to the level of being wrong.

We can start with mountain climbing.

IF SMART ALIENS were to observe our species, they would understand much of what we choose to do. Sexual intercourse, eating, drinking, resting, taking care of children, forming friendships—all of this is what one would expect from a creature that arose through natural selection. But they would be puzzled by much of what we've talked about so far in this book, like BDSM, horror movies, and marathon running. And at some point they would ask: What drives humans to do something as dangerous and difficult and seemingly useless as trying to climb Mount Everest?

Humans themselves don't know, either. The classic reply here is George Mallory's line, "Because it is there," which is funny because it's such a bad answer. All sorts of things are there, after all. This line is also the start of the title of an excellent article by the economist George Loewenstein, published about twenty years ago. The complete title is "Because It Is There: The Challenge of Mountaineering . . . for Utility Theory."

"Utility" is the technical term for the satisfaction you get from a good or service, and Loewenstein begins by noting that in the 1700s, at the time of Jeremy Bentham, there was a lot of interest in what, specifically, gave people satisfaction. As economics developed, though, interest in this psychological question waned, and now claims about the importance of utility in economic behavior end up providing, as Loewenstein puts it, "little more psychological insight than the observation that people choose what they prefer." One goal of his article was to resurrect interest within economics about the nature of utility, using mountain climbing as a case study.

If Bentham were around today, what would he have to say about the utility of mountain climbing? Although he is sometimes dismissed as a simple-minded hedonist, his actual views were sophisticated. He talks about sensory pleasure, but for him this was just one of many aspects of utility—he also considers more abstract forms of satisfaction, including (to use his own terms) the pleasures of skill, self-recommendation, a good name, power, piety, benevolence, and malevolence.

The utility of mountain climbing isn't obvious, to say the least. Sensory pleasure is a nonstarter. Loewenstein goes through reports about serious mountaineering (he includes also polar exploration in this category) and sums it up as "unrelenting misery from end to end." Diaries and journals by climbers talk about "relentless cold (often leading to frostbite and loss of extremities, or death), exhaustion, snow-blindness, altitude sickness, sleeplessness, squalid conditions, hunger, fear . . ." There is constant craving for food. And there is boredom: "On a typical ascent, the vast majority of time is spent in mind-bogglingly monotonous activities–for example, being 'weathered out' for many hours in a small smelly tent crammed in with other climbers." There is dread and anxiety, and rightfully so, given how many of the difficult expeditions end in death or mutilation.

In writing his article, Loewenstein drew upon reports from the climbers themselves, but there are also several excellent climbing documentaries and reenactments–such as *Into Thin Air*, *Everest*, *North Face*, and *Touching the Void*–and these support his point as well. Even if you stop before the part where everything goes pear-shaped and people die or lose their faces and toes, it's clear that nobody is, in any usual sense of the word, having "fun." Mountain climbing is a grim pursuit.

What about the pleasures of social connection, such as belongingness, friendship, and love? Some harsh activities do infuse the participants with the warmth of solidarity, providing a deep connection that perhaps one can get only through mutual struggle and suffering. Whatever one says about the horrors of war, social connectedness is a common positive theme–brothers-in-arms and all that. But this doesn't seem to be present in endurance climbing. Perhaps it's because the difficulty of breathing makes conversation harder, or maybe it's due to the constant physical stress, but climbers describe their experiences as lonely and alienating. There are stories of days and weeks spent in bitter silence, of disagreements that don't get smoothed over, of people parting with hatred.

A more promising candidate for the utility of mountain climbing is what Bentham called the pleasure of a good name. Years ago, I was watching my son compete in a climbing competition in a large gym in Connecticut and a crowd formed around a young woman who had walked in. She had just climbed Everest, and everyone wanted to hear her story. I can only imagine what the reaction would be if she were one of the first. One of the benefits of certain activities is the respect and admiration you get from others. This relates to difficulty and risk and ability in an obvious way. If climbing Everest were pleasant and easy, nobody would be impressed that you did it.

Prestige motivation can be awkward to talk about. Loewenstein notes that when climbers talk about their plans to conquer some mountain, they rarely admit that they are driven by a desire for fame. Similarly, expeditions such as Arctic treks are often described as being primarily scientific efforts or humanitarian projects, but Loewenstein cynically (and correctly, I think) sees this sort of framing as an attempt to mask other, less altruistic goals.

This is true for academics as well. Everyone who wins a big award in my field talks about how happy they are because it enables them to pursue their work and to support the research of their worthy students and colleagues. Still, our motivations are rarely pure. If you doubt this, note that, just as in mountaineering, science has bitter arguments over priority—over who got there first.

ANOTHER MOTIVATION FOR activities such as mountain climbing is curiosity about one's own capacities. Loewenstein notes that, just as with prestige, this motivation helps make sense of the difficulty that climbers endure: "Mountaineering reveals character only because it is not easy. A big part of the purpose of a trip is to test one's own mettle, and pain and discomfort provide the grist for such tests."

People really are curious about themselves. You'll know what I'm talking about if you ever took an online IQ test, or tried to figure out which of the sixteen personalities you have according to the Myers-Briggs test, or checked to see which Marvel Superhero you are. On Twitter, a writer I much respect and admire described taking an internet survey that promised to tell her real age based on the ingredients she preferred on her Subway sandwich.

"Know thyself" is good policy. If you are a coward, say, it's worthwhile to know it so that you can steer away from situations where bravery is at a premium. If you get tetchy when you don't have enough sleep or enough food, this is news you can use. But there are many things about yourself you just can't learn while sitting on your bum. Everyday life offers few opportunities to check out your capacity for bravery in the face of death or your tolerance for extreme physical challenges. If you want to know this about yourself, to test your mettle, activities such as mountain climbing seem like just the thing.

But this isn't as straightforward as it looks. If I want to see how strong I am, I can lift some weights to find out. But suppose I want to see how generous I am. Can I decide to volunteer at a homeless shelter and see how long I last? Not really—if I'm doing this to assess my kindness, then my time in the shelter will reflect my curiosity (or self-doubt), not my concern for the homeless. This paradox was nicely explored in the television program *The Good Place*, where the characters had to perform good deeds so as not to be forever condemned to hell but were in the unfortunate situation of knowing that their good deeds would redeem them, which meant that their motivations were tainted and couldn't really count as genuinely good deeds.

It's not just in acts of kindness that this issue arises. Picking a fight with the biggest guy in the room might look like a useful way to assess how tough I am, but if this is why I'm doing it, then I'm not tough at all—I'm painfully insecure. Similarly, choosing to mountain-climb in the service of mettle-testing might reflect your self-doubt, not your courage and taste for adventure.

Also, climbing Mount Everest is expensive, time-consuming, risky, and difficult. Is it really worth so much to learn something about oneself that's pretty irrelevant to one's life?

There is a related account, which has to do with the desire for reassurance, a wish to *impress* oneself. Call this self-signaling. But again, this seems like a lot of work for such a small benefit. Also, the whole idea is a bit confusing. Suppose you are confident that you can succeed at the climb. Great; now you don't need to do the climb in order to self-signal—you already know. Now suppose that you are not confident. Then it's lousy self-signaling—you would be engaging in a pursuit that might end up making you feel worse about yourself, not better.

Finally, imagine that someone climbed nearly to the top but, right before getting to the peak, had to turn back. Would the person be fully satisfied, since, after all, their efforts on the climb made plain their bravery, fortitude, and all that good stuff? Likely no; they would be disappointed. The goal is critically important.

THIS DESIRE FOR goal completion, to succeed, to bag that peak (as climbers put it), is Loewenstein's next possible explanation. Mountaineers talk about how the hunger to complete the climb can become overwhelming. At the start of their climb, they will often set a turn-back time, at which they have to begin to climb back down for the sake of safety, regardless of how far they've advanced. This is meant as a pre-commitment device, but as the peak draws near and it seems possible to achieve this goal that has taken so much expense and time and pain to pursue, many climbers find it impossible to follow their initial plan. Loewenstein points out that this refusal to turn back has led to many deaths, including, famously, seven on Everest in 1996.

This focus on goals does not deny the importance of prestige, or of signaling, to others and to oneself. The experience of mastery and flow, which we talked about in the last chapter, are surely relevant as well. But the main answer we've converged on is this: we climb Everest because it is the pursuit of the right sort of goal.

If we were to stop there, however, it wouldn't be much better than "Because it is there." What makes it the right sort of goal? One can respond with "Because it's meaningful," but this just pushes the question back. What makes for a meaningful goal? Why is climbing Everest meaningful but, say, walking up the stairs to my office is not? (It's not mere difficulty—walking up and down the stairs a thousand times would be difficult, but it seems more foolish than meaningful.) And what *is* meaning, anyway?

To inch up on an answer, consider some other pursuits that are also associated with meaning. You might not want to climb Everest, but maybe you want to go to war. Many people, and particularly many young men, do. There are obvious negatives here—the sacrifice, the separation from loved ones, the risk of mutilation or death. But it has an undeniable appeal.

People have different intuitions about going to war, and I can't resist a story here. I was once visiting a campus to give a talk and got picked up at the airport by someone from the philosophy department. We ended up, as one does, having a conversation about the morality of war. She had a son and talked about how upset she would be if he was drafted into battle and had to face combat. And I talked about my own sons and made the banal comment (or so I thought) that, whatever one thinks about war, there would be nothing worse for me than my sons' deaths. She responded by telling me she wasn't sure which would be worse—her son being killed or his having to kill another.

I was genuinely shocked by this. For me, having one of my sons get killed would be the worst thing imaginable, while hearing that my son killed someone would be . . . well, I don't know. It depends on the circumstances. Was he traumatized? What were the conditions of the killing? Was his act courageous, or gratuitous and cruel? Had he taken out a terrorist cell or been ordered to murder children? Actually, I told her, it was impossible for me to think of a scenario where my son being a killer would be worse than his death. The philosopher said she disagreed. She worried that killing would change her son in some terrible way. And, for her, perhaps that would be worse than his death.

She is right to note the transformative power of such an experience. But for many, this is part of war's appeal. Consider a controversial tweet by the famous author and prolific user of social media Joyce Carol Oates.

> All we hear of ISIS is puritanical & punitive; is there nothing celebratory & joyous? Or is query naive?

Predictably enough, her timeline was quickly filled with furious responses. People get upset when you discuss the charms of evil. But her query wasn't naive; it was sensible and important. As Ross Douthat noted in Oates's defense, "If you don't recognize that for at least some of the Islamic State's young volunteers there is a feeling of joy and celebration involved in joining up, then you're a very long way from understanding the caliphate's remarkable appeal."

Part of this appeal is the feeling of belonging. European recruits to ISIS are often recent immigrants; they are typically friendless and separated from family. There is a real hunger to be part of a community that embraces you, particularly if you have nobody else in your life.

But there's more. A group like ISIS offers the promise of a certain sort of transcendence, of suffering and pain and deprivation in the service of a greater good. My friend Graeme Wood, a journalist who has written extensively about ISIS, including a book based on interviews with both new recruits and long-standing members, tells me that many of those who joined the group were jaded when they signed up. They've had a lot of anonymous sex, they've taken every drug there is, they've lived lives of empty pleasure. But this wasn't enough. They were looking for more, something of real value.

I know it's unusual to talk about what Hitler got right, but here's George Orwell explaining the appeal of National Socialism in a review of *Mein Kampf*:

> Hitler . . . knows that human beings *don't* only want comfort, safety, short working-hours, hygiene, . . . they also, at least intermittently, want struggle and self-sacrifice. . . . Whereas Socialism, and even capitalism in

> a more grudging way, have said to people "I offer you a good time," Hitler has said to them "I offer you struggle, danger and death," and as a result a whole nation flings itself at his feet.

Now, very few people choose to join causes like ISIS, and in societies like ours, most don't even enlist in the military. But the appetite is vicariously enacted through fantasy and play. Cultural commentators often miss the powerful draw of video games, and particularly battle simulations. The *Call of Duty* series, for instance, sold about 400 million games, making $15 billion. Such simulations are popular because they scratch an itch.

Plainly, war has other appeals. I mentioned earlier the pleasure of belonging to a community, but there's also often a powerful moral pull: the desire to defend one's group and strike back at one's enemies. (This was particularly potent in the United States after the 9/11 attacks, when there was a sizable bump in enlistments.) Also, signing up for battle is an excellent signal of courage and loyalty, and, despite my qualifications above, it can serve as a way of learning about oneself. One can see a mix of these motives and others in a *New Yorker* interview with Adam Driver in which he talks about why he joined the Marines before becoming an actor:

> He craved a physical challenge, and the marines were tough. "They kind of got me with their whole 'We don't give you signing bonuses. We're the hardest branch of the armed forces. You're not going to get all this cushy shit that the Navy or the Army gives you. It's going to be *hard*.'" His decision to enlist was so abrupt

that a military recruiter asked if he was running from the law.

But war's appeal is more than belonging, morality, and signaling. As Chris Hedges put it in the title of one of his books, "War is a force that gives us meaning."

PERHAPS THE TWO examples so far have left you cold. Maybe you don't want to climb mountains *or* go to war. But what about having children?

There are few choices more important than this one, and psychologists and other social scientists have worked to figure out whether it's a good one. Much of this research suggests that, from a purely hedonic perspective, it isn't—having a child is a mistake. Research finds that the day-to-day experience of being a parent gives you little pleasure, particularly when your child or children are young. In one study, Daniel Kahneman and his colleagues got about nine hundred employed women to report, at the end of each day, each of their activities and how happy they were when they did them. It turned out that they recalled being with their children as less enjoyable than many other activities, such as watching TV, shopping, or preparing food. Other studies find that when a child is born, parents experience a decrease in happiness that doesn't go away for a long time, along with a drop in marital satisfaction that recovers only once the children leave the house. As Dan Gilbert puts it, "The only symptom of empty nest syndrome is increased smiling."

After all, having children, particularly when they are young, involves financial struggle, sleep deprivation, and stress. For mothers, there is also the physical strain of pregnancy and breastfeeding. And children can turn a cheerful and loving relationship into a zero-sum battle over who gets to sleep and rest and work and who doesn't. As Jennifer Senior notes, children provoke a couple's most frequent arguments—"more than money, more than work, more than in-laws, more than annoying personal habits, communication styles, leisure activities, commitment issues, bothersome friends, sex." Someone who doesn't understand this is welcome to spend a full day with an angry two-year-old (or a sullen fifteen-year-old) and find out.

But, as often happens in psychology, there are the initial studies that provide clear and interesting findings—such as "Having children makes you unhappy"—and then there are the later studies that find that it's more complicated. For one thing, the happiness hit is worse for some people than others. One study finds that older fathers actually get a happiness *boost*, while it's young parents and single parents, male and female, who suffer the greatest happiness loss. Also, most of the original data was from the United States. A recent paper looked at the happiness levels of people with and without children in twenty-two countries. They found that the extent to which children make you happy is influenced by whether there are childcare policies such as paid parental leave. Parents from Norway and Hungary, for instance, are happier than childless couples—while parents from Australia and Great Britain are less happy. The country with the greatest happiness drop when you have children? The United States.

Children make some people happy, others miserable, and the rest somewhere in between—it depends, among other things, on how old you are, whether you are a mother or a father, and where you are living. But there remains a deep puzzle. There are many people who would have had happier lives, and happier marriages, if they chose not to have kids. Yet they still describe parenthood as central to their lives, the best thing they've ever done. Why don't we regret our children more?

One possibility is that it's memory distortion. When we gauge our own previous experiences, we tend to remember the peaks and forget the 99 percent of mundane awfulness in between. Our memory is selective. Drawing on a distinction we made earlier in the book, Jennifer Senior puts it like this: "Our experiencing selves tell researchers that we prefer doing the dishes—or napping, or shopping, or answering emails—to spending time with our kids. . . . But our remembering selves tell researchers that no one—and nothing—provides us with so much joy as our children. It may not be the happiness we live day to day, but it's the happiness we think about, the happiness we summon and remember, the stuff that makes up our life-tales."

These are plausible enough ideas, and I don't reject them. But there are two other explanations for why people usually don't regret being parents that I want to talk about, and I'm going to get pluralist again, because neither of these has to do with happiness in any simple sense.

The first involves attachment. Most parents love their children, and it seems terrible to admit to yourself and others that the world would be better if someone you loved didn't exist. More than that, it's not just that you feel compelled to say that you are happy they exist—you *are* happy they exist. After all, you love them.

This can put one in the interesting situation of desiring a state that you believe would make you less happy than the alternative. In his book *Midlife,* Kieran Setiya expands on this point. Modifying an example from Derek Parfit, he asks you to imagine being in a situation where, if you and your partner were to conceive a child during a time frame, the child would have a serious, though not fatal, medical problem, such as chronic joint pain. If you wait, the child will grow up without such a problem. For whatever reason, you choose not to wait. The child grows up and you love him and, though he suffers, he is happy to be alive. Do you regret your decision?

That's a complicated question. Of course it would have been better to have a child without this condition. But if you waited, you'd have a different child, and this baby (then boy, then man) that you love wouldn't exist. It was a mistake, yes, but perhaps a mistake that you don't regret. The attachment one has to an individual can override an overall decrease in the quality of a person's life, and so the love we usually have toward our children means that our choice has value above and beyond whatever effect they have on our happiness.

A second, perhaps related, consideration is that psychologists and parents are talking past one another. When I say that raising my sons is the best thing I've ever done, I'm not saying that they gave me pleasure in any simple day-to-day sense, and I'm not saying that they were good for my marriage. I'm talking about something deeper, having to do with satisfaction, purpose, and meaning.

It's not just me. When you ask people, "How often, if at all, do you think about the meaning and purpose of life?" or "In the bigger picture of your life, how personally significant and meaningful to you is what you are doing at the moment?," parents—both mothers and fathers—say that their lives have more meaning than those of non-parents. And in the study by Baumeister and his colleagues on meaning and happiness, discussed at length earlier, they found that the more time people spent taking care of children, the more meaningful their lives were—even though they reported that their lives were no happier.

Just like mountain climbing and going to war, then, raising children is an activity that has an uncertain connection to pleasure but has the potential to enhance meaning and purpose. The writer Zadie Smith puts it better than I ever could, describing having a child as "a strange admixture of terror, pain, and delight."

Smith, along with everyone else who has thought seriously about these issues, points out the risk of close attachments: "Isn't it bad enough that the beloved, with whom you have experienced genuine joy, will eventually be lost to you? Why add to this nightmare the child, whose loss, if it ever happened, would mean nothing less than total annihilation?" But if the loss of a child would be a total annihilation, then having a child, healthy and happy and sound, must be annihilation's opposite, which sounds pretty terrific.

WITH THESE EXAMPLES of meaningful activities behind us, we can ask what they have in common, and how this relates to suffering. But before doing so, we need to make an important distinction.

Albert Camus wrote, in *The Myth of Sisyphus*, that there is only one serious philosophical problem, and this is whether to kill oneself. For Camus, what makes this a philosophical problem is that it connects to the issue of whether life is worth living, and this, in turn, connects to the meaning of life. Which makes the meaning of life the most urgent of questions.

But Camus is not saying, or at least shouldn't be saying, that people can bear to live only if they have an *explicit* answer to the question of the meaning of life. I have some older family members who have lived, and are living, rich lives, and if you asked them about the meaning of life, they would snort and roll their eyes. You can have a great life and never think about this question.

Not everyone agrees with me on this. For some philosophers, questioning the meaning of life is required in order to have a meaningful life. Casey Woodling writes, "One discovers this meaning or significance by evaluating one's life and meditating on it; by taking a step back from the everyday and thinking about one's life in a different way. If one doesn't do this, then one's life has no meaning or significance. . . . This comes close to Socrates' famous saying that the unexamined life is not worth living. I would venture to say that the unexamined life has no meaning."

Now, I'm all for introspection, but this goes too far. Consider two people. Jane is engaged in difficult and important projects. She has a large family and network of friends; she works to make the world a better place. But she never takes the time to examine her life. Maybe she is just too busy. By contrast, Moira lives on money left to her by her parents; she spends her days drinking, smoking weed, and surfing the web. But, between YouTube videos and hate-tweeting, she spends hours thinking about her life, mulling over its value, meditating on it.

Woodling would have to say that Jane's life, involving activities that most would see as meaningful and significant, has less meaning than Moira's. And this, to me at least, doesn't ring true. A meaningful life, at least to some extent, has to do with what one does and how one affects people.

So I'm not as prone as some scholars to worry about how much people mull over the meaning of life. Emily Esfahani Smith talks about the American Freshman Survey, which found that in the late 1960s, 86 percent of respondents claimed that "developing a meaningful life philosophy" was "essential" or "very important," while in the 2000s, the proportion dropped to 40 percent. She is disappointed in this; she sees it as a bad sign. I don't. It might reflect a lack of interest in a meaningful life, and less engagement in one. But it could also be that the youth of today are less full of themselves. Maybe they put more energy into living and less into thinking about how they are living. I have a lot of philosopher friends, some of whom think all the time about deep questions of meaning and purpose, and while I do love philosophers, they don't seem to be better people than the non-philosophers I know, and I'm not sure that there's any interesting sense in which their lives are more meaningful than everyone else's.

The general point here is that you can achieve a meaningful life without knowing that you're trying to achieve it or thinking about it at all. To give you an analogy, suppose I was writing a book in favor of physical exercise. It's tempting to spend one's free time surfing the web and eating snacks, but getting up and running or biking or weight-lifting, though difficult, leads to all sorts of long-term benefits, physical and psychological. (I think, by the way, that all of this is true.) To make my case, I would cite studies that find that people who exercise are healthier than those who don't. Exercise is good for you.

But none of this would imply that people who exercise need to have an explicit theory of exercise. Maybe they don't know what they're doing is good for them, or they don't think of their favorite activities as exercise at all, or never give it any thought.

This is surely true for young children, who are often active and energetic but don't reflect on the value of their activities.

So, too, with meaning. Some people engage in meaningful pursuits, and this, I argue, makes their lives better. But people don't have to think about meaning for this to work. People who mountain-climb, for instance, might have an entirely mistaken theory of what climbing does for them, just as someone who exercises might have an entirely wrong theory of the benefits of exercise.

WE ARE TALKING about meaning and meaningful pursuits and meaningful lives. But if you're looking for an answer to the question "What is the meaning of life?" you've picked the wrong book. I'm comfortable talking about a meaningful life, so long as we're just talking about a life full of meaningful pursuits and meaningful experiences. But I don't think there's some singular answer to this often asked question.

I'm following here Douglas Adams, who, in *The Hitchhiker's Guide to the Galaxy*, tells the story of how, millions of years ago, alien scientists constructed a computer to provide "the answer to the Ultimate Question of Life, the Universe and Everything." Finally, the computer arrives at the answer: forty-two. The scientists are outraged.

> "Forty-two!" yelled Loonquawl. "Is that all you've got to show for seven and a half million years' work?"
>
> "I checked it very thoroughly," said the computer, "and that quite definitely is the answer. I think the problem, to be quite honest with you, is that you've never actually known what the question is."

"But it was the Great Question! The Ultimate Question of Life, the Universe and Everything," howled Loonquawl.

"Yes," said Deep Thought with the air of one who suffers fools gladly. "But what actually *is* it?"

A slow stupefied silence crept over the men as they stared at the computer and then at each other.

"Well, you know, it's just Everything . . . everything . . ." offered Phouchg weakly.

"Exactly!" said Deep Thought. "So once you know what the question actually is, you'll know what the answer means."

In Adams's story, the scientists then decide to build another computer to make sense of the question. But this is where I part company with Adams: I don't see this as a question that needs clarity. Rather, it's like "Where does your lap go when you stand up?" Or, from Ludwig Wittgenstein, "What time is it on the sun?" They're not good questions to ask.

If you try to answer a dumb question, you'll get an unsatisfying answer. The philosopher Tim Bale states, "The meaning of life is not being dead." He suggests that just living is enough for a life to have meaning, like a hippie professor who gives everyone an A just for showing up. I can't imagine that someone would be satisfied by such an answer; it's not much better than forty-two. The same sort of concern holds for the many philosophers who say that there is no meaning of life. For me, this is like asking whether a bicycle suffers from clinical depression and answering that it doesn't. There is a sense in which this is correct, but a better answer would be that the notion of being clinically depressed just doesn't apply to bicycles.

Bicycles aren't the sorts of things that get depressed, just as life isn't the sort of thing that has a meaning.

I do need to make one qualification here: there are some people for whom the question "What is the meaning of life?" can be sensible. Think of how we use the word "meaning." If you talk about the meaning of a sentence, or of a weird symbol drawn in the sand, or a cryptic email, you often are talking about intention. From that perspective, someone who believes that we are the creations of God can talk sensibly about the meaning of life, because this refers to God's intention for us, his plan. And, indeed, if you ask some people about the meaning of life, they will tell you to read the Bible, the Torah, or the Quran and say, sincerely, that it's all in there.

Absent a supernatural creator, though, we have to give up on the question "What's the meaning of life?" This position is nicely expressed by Viktor Frankl:

> To put the question in general terms would be comparable to the question posed to a chess champion: "Tell me, Master, what is the best move in the world?" There simply is no such thing as the best or even a good move apart from a particular situation in a game and the particular personality of one's opponent. The same holds for human existence. One should not search for an abstract meaning of life. Everyone has his own specific vocation or mission in life to carry out a concrete assignment which demands fulfillment.

All of us have the sense that some activities are meaningful and others are not. In our discussion of psychological research throughout this book, we've seen that people are capable of rating activities and experiences based on their meaning, and that they have clear feelings about how much meaning their lives have. Many people would agree that helping the poor adds meaning to their life, while, say, binge-watching Netflix or smoking pot all the time doesn't. We can also talk sensibly about meaningful experiences, such as the birth of a child, as opposed to less meaningful experiences, like discovering that there is an extra doughnut in the box when you thought you had eaten them all.

The ability to make such distinctions gives us a strategy to pursue. As an analogy, consider the study of consciousness, a messy and complex topic. Researchers can get some purchase on it by asking about the difference between situations in which someone is conscious, like presumably right now for you, and situations in which someone is not conscious, like when in a coma. Or between experiences that one is conscious of, like reading this book, and those that one is not, such as the feeling of your feet in your shoes or on the floor. (Right now the experience is conscious, but it probably wasn't two seconds ago.)

The same strategy can work if we want to capture our everyday intuitions about what meaning is. We can ask: What distinguishes a meaningful activity or meaningful experience from one without meaning? And many people have done exactly that.

Emily Esfahani Smith talks about how in 1932, the historian and philosopher Will Durant published a book called *On the Meaning of Life*. In it, he compiled answers he got from various luminaries of his time, such as Mohandas Gandhi, Mary

Woolley, and H. L. Mencken. About fifty years ago, *Life* magazine did the same thing, writing to more than a hundred influential people—such as the Dalai Lama, Rosa Parks, Dr. Ruth, John Updike, Betty Friedan, and Richard Nixon—and asking them the same question. Now, as I have just argued, we should be suspicious of that precise question ("meaning of life"), but most of the answers actually were often about meaningful *activities*, which is closer to what we're looking for.

Smith discusses the responses and summarizes them as follows:

> Each of the responses to Durant's letter and *Life*'s survey was distinct, reflecting the unique values, experiences, and personalities of the respondents. Yet there were some themes that emerged again and again. When people explain what makes their lives meaningful, they describe connecting to and bonding with other people in positive ways. They discuss finding something worthwhile to do with their time. They mention creating narratives that help them understand themselves and the world. They talk about mystical experiences of self-loss.

Smith organizes her own book, *The Power of Meaning*, around four themes that show up in this summary:

Belonging: connecting to and bonding with other people
Purpose: finding something worthwhile
Storytelling: narratives that bring order to life
Transcendence: mystical experiences of self-loss

There are other, related proposals. In several papers, based on a review of the psychology literature, Michael Steger talks about three features of meaningful activities; these are similar to Smith's:

Coherence: making sense, fitting into a narrative
Purpose: directed toward a goal
Significance: worthwhile, having value, and importance

In an article called "Beyond Bentham: The Search for Meaning," George Loewenstein and Niklas Karlsson present their own list of features, which include:

A resolution of purpose or goals: figuring out what you are aspiring toward
An expansion of the self through time or across persons: binding oneself to a broader group of people or to past and future generations
An interpretation of one's life: the creation of a narrative of one's life

These overlap nicely with the other two lists of criteria.

Before I take a shot at my own definition, let's step back and remind ourselves what we're doing here. It's not like we're studying meaning independent of how people think of it, as if we have found a strange animal in the woods and we're speculating about its properties. Rather, this is a form of conceptual inquiry. We know people have a sense of meaning; we are trying to unpack it and see what properties it has. We want to do this because meaningful pursuits and meaningful events, as people define them, have real value.

Here is my own attempt to integrate the ideas so far, looking first at meaningful activities.

> A meaningful activity is oriented toward a goal, one that, if accomplished, would have an impact on the world—and this usually means that it has an impact on other people. This activity extends across a significant portion of one's life and has some structure—it's the sort of thing that one can tell a story about. It often connects to religion and spirituality and often connects to flow (leading to the experience of self-loss) and often brings you into close contact with other people and is often seen as morally virtuous—but none of these additional features are essential.

I agree here with Smith about the importance of transcendence—or, put differently, spirituality and religion—in meaningful pursuits. It is important enough that I will devote much of the next chapter to it. But it's not essential. Meaning is available to those who disavow the transcendent. Someone who climbs Everest or raises foster children or gives their life to fight the Empire might be a stone-cold atheist, unmoved by any spiritual beliefs. At the start of this book, I cited this tweet by Greta Thunberg:

> Before I started school striking I had no energy, no friends and I didn't speak to anyone. I just sat alone at home, with an eating disorder. All of that is gone now, since I have found a meaning, in a world that sometimes seems shallow and meaningless to so many people.

I don't know whether Thunberg would say that this had a spiritual element, but it's certainly possible that she would not. The human significance of her project—its felt importance—is sufficient to make it meaningful.

Another one of Smith's criteria is belonging, and I do agree that most meaningful activities bring you into contact with other people. This is also related to the Loewenstein and Karlsson notion of "expansion of the self." But, again, this is not essential. There are meaningful pursuits that are solitary. Alex Honnold's free-solo climb of El Capitan is a good example of this. He was followed by a film crew, but this wasn't a necessary part of the pursuit, and Honnold gives every impression of being someone who prefers to work alone. Another example is that of Andrew Wiles, who spent years in solitude solving Fermat's Last Theorem. For many, including me, this is a project of great meaning, but it wasn't a particularly social one.

What about morality? Many meaningful pursuits are moral. But consider again Adolf Eichmann, the architect of Hitler's Final Solution, who was by all accounts engaged in a project that had a profound impact on the lives of millions—a profoundly terrible impact. While we might not like to call what he was doing "meaningful" (since this seems like a positive term), I don't see a principled way of excluding it. To use a milder example, climbing Mount Everest is seen by many as worthwhile and important, but I doubt that even the climbers themselves would frame it as *good*.

Like most analyses, my conception of a meaningful activity is centered around significance and impact. Such notions are inherently vague. There are some actions that almost everyone would agree are not meaningful—eating a cookie, say—and others that almost everyone would see as making the cut, such as devoting your life to ending world hunger. But a lot falls in between, and activities that would be meaningful to one person might not be thought of this way by another. We can double back now to the examples we discussed earlier, of mountain climbing, war, and raising children, and we can see how they fit our criteria. They all involve an impactful goal (this is hardest to see in the case of mountain climbing, but certainly the climbers themselves see it as impactful). They extend over a long period of time and involve a series of events; they have a narrative structure. When it comes to the optional criteria, they match some but not others: they are all social, and some are seen as morally valuable (again, mountain climbing the least of all), but, while religion can infuse all of these activities, it doesn't have to.

You'll notice that suffering is not one of the criteria here. But given that meaning involves the pursuit of significant and impactful goals, meaning will inevitably come with suffering—with difficulty and anxiety and conflict and perhaps much more. When one chooses to have a child or go to war or climb a mountain, one might not wish for or welcome suffering. But it always comes along for the ride.

WE'VE BEEN TALKING about meaningful pursuits, but there are meaningful experiences as well. Here the bar is dropped somewhat. These can be more passive and don't necessarily involve achieving a goal. What seems to be key here is that they will change you in some way. This change can be profound, as in giving birth, or it can just be distinctive and memorable in a smaller sense, the sort of thing that would make for a good story. As with meaningful activities, what makes for a meaningful experience is a matter of degree.

In a recent paper, people were asked to think back on the most significant experience (in one study, over the past year; in another, over the past three months), to describe the experience in a paragraph, and then to rank how meaningful it was, from 0 ("a meaningless experience") to 10 (either "the most meaningful experience you can think of" or, in another study, "the most meaningful experience you can imagine anyone having"). They were also asked to indicate the extent to which the experience was pleasurable or painful.

It turned out that the most meaningful events tended to be on the extremes—those that were very pleasant or very painful. These are the ones that matter, that leave a mark.

Consistent with this, in his discussion of mountain climbing, George Loewenstein notes that some of the people who went through the worst ordeals were the most positive about their experience. Maurice Herzog, who in 1950 was part of the first team to summit Nepal's Annapurna, lost several fingers and parts of his feet but, on the plus side, said that the ordeal "has given me the assurance and serenity of a man who has fulfilled himself. It has given me the rare joy of loving that which I used to despise. A new and splendid life has opened out before me." Beck Weathers, who lost his hands and much of his face after being left overnight in a blizzard on Everest, says, "I traded my hands for my family, and it is a bargain I readily accept." Loewenstein drily concludes, "Meaning-making may also be enhanced by the loss of body parts."

Now, nobody chooses extreme negative events such as those experienced by these mountain climbers. But we often do seek out more minor negative experiences, in part for their transformative natures but also because we might simply want to possess these experiences later on. We want to store them in memory and, to use a strange but apt term, consume them in the future. To do this, we're willing to suffer, or at least to forgo pleasure. As Seneca put it, "Things that were hard to bear are sweet to remember."

This is explored in an interesting series of studies. In one of them, people got a series of questions like these:

> You have a six-hour layover in the Budapest airport. Would you rather stay at the airport and watch DVDs on your laptop computer or explore the city in extremely cold weather?
>
> You're on vacation. Would you stay at a Marriott hotel in Florida or at an ice hotel in Quebec?

For each of the two options, subjects were asked, "Which would be more memorable?," "Which would be more pleasant?," and "Which would you choose?" It turned out that the chosen option was typically the most memorable one, even though for every choice pair it was also rated as the least pleasant. Most people said that they would choose to explore Budapest or go to the ice hotel, even though most of them also predicted that staying in the airport and going to Florida would be more pleasant.

Another study zoomed in on the Florida/Quebec contrast with a new set of subjects, this time asking them to explain their choices. About one-third chose the Florida vacation, and they often described their choice as based on pleasure, using words such as "fun," "enjoyable," or "pleasant." Such words were rarely used by the ice-hotel choosers, who instead often talked about this choice as a way to acquire new memories, describing such a vacation as "challenging but still a memory maker" and as "cold, new, and memorable."

A third study was done in Times Square on New Year's Eve. The researchers tested people who had already been standing outside for hours in freezing weather. Some people were prompted to think about the immediate moment ("Are you happy right now about your choice to come to Times Square tonight?"); others were motivated to think about the future ("Ten years from now, when you will look back at your choice to come to Times Square tonight, do you think you will be happy about your choice to come to Times Square?"). Then they were all told, "Tonight it is expected to snow in New York," and were asked whether they hoped the sky would be clear at midnight or whether they hoped it would snow.

People were more likely to prefer snow when motivated to think about the distant future rather than the present. They were also more likely to prefer snow when they were told, "This would be the first time in the last 15 years that it would be snowing on New Year's Eve, in New York, at midnight"—presumably because this would make it a special and collectible experience.

These are all examples of how we can choose meaningful experiences so as to store them up for the future. But you can also seek out meaningful experiences in the past. Viktor Frankl described how he and his fellow prisoners were being punished through starvation, and he worried that some would commit suicide. So he talked to them about the present (*It could be worse*) and about the future (*It will be better*). But this was not all.

I did not only talk of the future and the veil which was drawn over it. I also mentioned the past; all its joys, and how its light shone even in the present darkness. Again I quoted a poet—to avoid sounding like a preacher myself—who had written, "*Was Du erlebst, kann keine Macht der Welt Dir rauben.*" (What you have experienced, no power on earth can take from you.) Not only our experiences, but all we have done, whatever great thoughts we may have had, and all we have suffered, all this is not lost, though it is past; we have brought it into being. Having been is also a kind of being, and perhaps the surest kind.

6

SACRIFICE

In a Hindu festival in Mauritius, celebrants walk on hot coals and have skewers plunged through their cheeks and tongues. They have hooks dug into their backs and stomachs, and these hooks are attached to chariots that weigh hundreds of pounds. The men then spend hours in the afternoon heat dragging these chariots to the top of a distant hill.

Such an ordeal is extreme, but there are lesser forms of chosen suffering in other religions, such as the denials of pleasure present in Lent, Yom Kippur, and Ramadan. And all religions have restrictions and sacrifices that apply 24/7/365. These involve what you can eat and who you can have sex with (and when, and how), but they are also of broader scope—as someone who spent his teen years going to an Orthodox Jewish synagogue, I'm aware that the rules of the Torah constrain *everything.* Religions also preach what they practice; holy texts are explicit about the importance of rejecting worldly pleasures and the virtues of experiencing sacrifice and pain.

Now, if you've made it this far into the book, you'll know that similar sorts of chosen suffering show up in secular contexts. The agonies of the celebrants in Mauritius don't look all that different from extreme forms of BDSM. Sacrifice and deprivation? Long periods of stillness and silence? You see these in athletic training and meditation. Abstract arguments for the importance of suffering as the key to the good life? There are plenty of nonreligious doctrines that say the same thing.

But no discussion of chosen suffering would be complete without looking at religion. Religions exemplify the social value of suffering—not as a cry for help, as in the self-harm inflicted by troubled adolescents, but as social glue, the sort of belongingness that we discussed briefly in the last chapter as one feature of meaning. Perhaps more important, religion provides our species' longest and deepest struggle to make sense of suffering, including suffering that is unchosen.

RITUALS ARE PART of all religions. Some are painful, like scarification and circumcision. Others are harmless or even pleasurable, involving singing and dancing, body painting, and communal feasts.

It's long been a puzzle why rituals exist. A lot of people, including me, think that the psychological foundations of religion are part of human nature. But this can't be right for specific rituals. No two-year-old has instinctively turned to Mecca to pray or said the Hebrew prayer for bread, and none ever will. Rituals are cultural inventions.

A more promising approach to explaining them, then, appeals to cultural evolution. Just as natural selection occurs because certain constellations of genes help some animals survive and reproduce more than others, cultural evolution occurs because certain social practices help some communities outlast and outgrow the rest. If society A has practice X and society B doesn't, and if, because of practice X, society A does better, then you are more likely to see society A—along with practice X—in a hundred or a thousand years.

One useful sort of practice unites people. Societies flourish when the members of the group are willing to tone down their selfish motivations and care for those around them. This is said to be one function of religion more generally—as Jonathan Haidt puts it, "Religions . . . work to suppress our inner chimp and bring out our inner bee," releasing our hive morality, in which the group is all that matters. One way religions do this is through ritual. And this does seem to do the trick: if you want to know which societies will last the longest, a strong predictor is the number of hours a day that they engage in ritualized behavior.

How does this work? For one thing, rituals can generate what Émile Durkheim called "collective effervescence." Think about the practice of linking arms and dancing at a Jewish wedding, for instance. There is substantial evidence that this sort of synchrony brings people together and makes them care for one another more.

But not all rituals have this feature of joint synchronized behavior—painful ones typically do not. Usually, only some people choose to experience the pain; most are just spectators. But such rituals can generate a different sort of connection, one grounded in empathy. Consider again the Hindu celebrants of Mauritius. Dimitris Xygalatas finds that those who engage in high-pain rituals become more loving to their group and more generous as a result. And the more pain they experience, the more group-oriented they get. Importantly, this growing attachment to the group holds true not only for the participants themselves but also for those who watch their performance, who watch them on the long trek up the hill. These observers report feeling vicarious pain, and this brings them closer to their community.

Rituals might also provide some benefit to the individuals who engage in them, not just to the groups to which they belong. This is particularly true for painful rituals: choosing to engage in them can signal commitment to the group and display courage and virtue. Xygalatas points out that those who ask to have the most skewers put into them and who pull the most weight (literally) are young men looking for romantic partners. They also tend to be relatively poor, because if you're rich you have easier ways to show off your value.

Xygalatas notes as well the risk involved in this sort of ritual. The man gets to choose how many skewers and how much weight to pull. The more the better, of course, but if he chooses too much, he might fail and not make it up the hill, and this is a social disaster—it's a very public sign of weakness and, worse, it's taken as evidence that the gods do not think well of you.

This is, after all, how the participants themselves think of what they are doing—in terms of what the gods think. Rituals are not typically seen by the participants as mechanisms of group stability and signaling. Those who fast for Yom Kippur or give up sweets for Lent see themselves as following the commands in holy texts and the wishes of God or, for the less devout, simply following tradition or honoring family obligations. Rituals work best, it seems, when their functions are obscured. It's hard to imagine the Mauritius ritual or the Passover Seder persisting in a culture where everyone was explicit that this was a mechanism to increase group solidarity and there was nothing else to it. If this is right, then self-conscious attempts to use rituals to build communities are destined to fail—rituals won't work if they are done for utilitarian ends.

One might even suggest that we invented gods to give us a rationale for performing rituals, like painting a bull's-eye on a tree to motivate us to practice our archery. But this is too strong. You don't need God or religion to have rituals, including painful rituals. In some Brazilian jiu-jitsu clubs, for instance, those promoted to a higher level are run through a gauntlet of other club members and whipped by belts, leaving their backs and necks raw and bloody. This is seen as an important, even transcendent, experience.

Occasionally, a public masochistic display similar to what you see in the Hindu celebrants is created de novo—not a ritual at all, but an inspired act of chosen suffering to impress and show allegiance to the group that you belong to. My son told me of an event he went to for a skiing-and-snowboarding club at his university. At this event, certain leadership positions were competed for. One by one, the candidates went onstage in front of the membership of the group, spoke about which position they were competing for, and then did something that's intended to impress—telling a joke, doing backflips, that sort of thing. At the event my son attended, a young man went up and said that he didn't really want any of the positions, but he was wondering if it would be all right with the audience if he did something anyway. People agreed.

And then he reached into a bag and took out six mousetraps. He snapped them down on every finger on his left hand

and then his tongue. And then he pulled out a bottle of hot sauce and squirted some into his right eye and then his left eye. Finally, he took out a sign that said UBC SKI AND BOARD and a stapler and *he stapled the sign to his chest*.

The crowd stood up and cheered, and the club authorities created a new leadership position in his honor.

MY BOOK HAS been a discussion and defense of chosen suffering, and there's more to come. But what about unchosen suffering? What about the stuff that you don't want? A long wait at the DMV, stubbing your toe, severe back pain, losing your home to a tsunami, the death of a child, being tortured, spending years in a concentration camp. Not suffering that you've opted for; not suffering that's the by-product of a chosen meaningful pursuit or that reflects a social commitment or moral decision; not suffering that you can say "stop" to at any point. Suffering that happens whether you want it or not.

Take James Costello. In April 2013, he was cheering on a friend near the finish line of the Boston Marathon when the bombs exploded. Costello was sprayed with shrapnel, and his arms and legs were severely burned. He went through months of surgery and rehabilitation. If we ended here, it would be simply an illustration of the obvious—bad things happen, and we need to recover from them.

But Costello's story had a twist. While in the hospital, he fell in a love with a nurse, Krista D'Agostino, and they became engaged. And then Costello posted a picture of the ring on Facebook and wrote, "I now realize why I was involved in the tragedy. It was to meet my best friend, and the love of my life."

It's not an unusual reaction to unchosen suffering. As the

phrase goes, *everything happens for a reason.* Google the phrase, look it up on social media—it's everywhere. Maybe you've said it once or twice yourself.

And perhaps it's a good thought to have. Daniel Gilbert talks about the "psychological immune system," a part of our psyche that recovers from negative experiences by giving them meaning. He tells the story of Katie Beers, who, when she was nine years old, was abducted, raped, and tortured for two weeks in the dungeon of a family acquaintance. Her description of the event twenty years later? "The best thing that happened to me." Or take Moreese Bickham's description of spending thirty-seven years in the Louisiana State Penitentiary for a crime he didn't commit: "I don't have one moment's regret. It was a glorious experience." Or recall, from the last chapter, George Loewenstein's reports of terrible climbs, of the man who lost several fingers and parts of his feet and said that "a new and splendid life has opened out before me." Or another who said, "I traded my hands for my family, and it is a bargain I readily accept."

Once you look for this, you see it everywhere. My friend and colleague Laurie Santos has an excellent podcast on happiness, and in one episode she interviews a young man named J. R. Martinez who served in Iraq and was trapped in a flaming jeep that was blown up by an IED. He was hospitalized for a long period and maimed for life. As he tells his story, with its awful details (such as when he sees his scarred face in the mirror for the first time), you can see it coming.

Santos: Would you change anything? Would you do it over differently?

Martinez: Nah, I wouldn't change anything. I one hundred percent mean that.

Santos: But you wouldn't change the explosion, the scars, the surgery? You would keep all of that?
Martinez: Yeah . . . I'm blessed.

Such examples are vivid. But are they representative? In a series of papers written a few years ago, my then-graduate student Konika Banerjee and I explored how common it is to give meaning to various life events.

In one study, we asked people to first reflect on significant events from their own lives, such as graduations, the births of children, falling in love, the deaths of close family members, and serious illnesses. We asked them whether they felt that these events were caused by fate, whether they were meant to be, whether they happened for a reason, and whether they happened in order to send them a message. We found that people often answered yes to all or most of these questions, for the negative events as well as for the positive ones, and that this was often true even for self-described atheists. In other work, we found that even young children show a bias for believing that life events happen for a reason—to "send a sign" or "to teach a lesson"—and that they do so to a greater extent than adults.

These findings were interesting to us because they suggest that belief in fate and karma might be universal. But we also found that religion had a large impact on how people viewed these events. For instance, when we asked participants whether they believe, for significant life events, that they "happened to send me a message," religious people were more than twice as likely to agree. Similar differences occurred for questions about whether something was "meant to be" or "happened for a reason."

It's not surprising that religion has this effect. Just as religion can provide a coherent answer for the meaning of life (as discussed in the last chapter), it can also, and relatedly, make sense of unchosen suffering. It does so in multiple ways.

Some religions teach that suffering is the product of a beneficent form of discipline. God is a stern father who loves you, and so the hardship you endure is the punishment he inflicts—but he does so for your own good. This is explicit in the New Testament:

> Endure hardship as discipline; God is treating you as his children. For what children are not disciplined by their father? If you are not disciplined—and everyone undergoes discipline—then you are not legitimate, not true sons and daughters at all. Moreover, we have all had human fathers who disciplined us and we respected them for it. How much more should we submit to the Father of spirits and live! They disciplined us for a little while as they thought best; but God disciplines us for our good, in order that we may share in his holiness. No discipline seems pleasant at the time, but painful. Later on, however, it produces a harvest of righteousness and peace for those who have been trained by it.

This passage ends on a positive note, with the promise that this discipline will serve to make our lives better, producing "a harvest of righteousness and peace." But the earlier passages are less results-based, focusing on *legitimacy*—without such punishment, we would not be "true sons or daughters at all," and so, by being punished, "we may share in his holiness." Here the discipline is seen as part and parcel of our relationship with God. After all, good fathers discipline their children, so our suffering—even putting aside any positive effects that it might have—reflects the love that God has for us.

A different explanation of suffering in the Christian tradition involves one's relationship with Christ. One extreme example can be found in the Philippines, where, on Good Friday, Catholic penitents have themselves nailed to crosses—crucified. But it applies to unchosen suffering as well. One commentary sums up the teachings of Saint Paul: "Because we are being saved through the death and resurrection of Christ we must participate in his Passion to obtain salvation." This theme is explored as well by Pope John Paul II, who writes that, by sharing the suffering of Christ, the devout "in a sense, pay back the boundless price of our redemption."

Yet another approach to the utility of suffering is presented by C. S. Lewis. He worries that we get too complacent and proud in our happiness; suffering wakes us up. As he puts it, in characteristically beautiful prose, "But pain insists upon being attended to. God whispers to us in our pleasures, speaks in our conscience, but shouts in our pains: it is His megaphone to rouse a deaf world. . . . It removes the veil; it plants the flag of truth within the fortress of the rebel soul."

One might see these explanations of suffering as just something that religions tend to provide, in the same way that they tend to contain theories of the cause of mental illness or dreams. But perhaps these explanations are more important than this. When cognitive scientists talk about the functions of religion, we often say that it satisfies our curiosity about certain big questions—it tells us where the universe came from, how humans and animals came to be. But it's not clear that these broad metaphysical issues are what we most worry about. I think I speak for most people when I say that if I never have a good theory of the origin of the universe, I'll do just fine. But the need to make sense of suffering is more urgent, particularly when we ourselves are suffering. We wish to be reassured, to know that it's not all in vain, to hear that, maybe, our suffering will stop and we will be rewarded, whether in the future on earth, in heaven, or in another life. Religion's idea that suffering is good is a message that we are eager to hear.

This is illustrated in a short story by Ted Chiang, called "Omphalos," where people from an alternative universe discover that their world was likely created as a discarded trial run before the real creation—that they are not the object of God's love. The narrator describes the effect that this discovery had on one man:

> Dr. McCullough said, "You are childless, so you can't comprehend the pain caused by losing a son."
>
> I told him he was correct and said that now I realized why this discovery must be especially difficult for the two of them.
>
> "Do you really?" he asked.
>
> I told him what I surmised: that the only thing that had made his son's death bearable was the knowledge it was part of a greater plan. But if humanity is not in fact the focus of your attention, Lord, then there is no such plan, and the son's death was meaningless.

In a less secular time, the value of unchosen suffering was more broadly accepted. In the early 1800s, anesthesia, including nitrous oxide and ether, was developed. To us moderns, this seems like an unmitigated good—if you ever wish you lived in the past, just read some descriptions of surgery in a pre-anesthesia world. When P. J. O'Rourke was challenged on what's so good about modern times, he supposedly instantly responded, "Dentistry!"

But many at the time thought anesthesia was an abomination. William Henry Atkinson, the first president of the American Dental Association, wrote, "I wish there were no such thing as anesthesia! I do not think men should be prevented from passing through what God intended them to endure."

I find this ridiculous, and I figure you do as well. But it's not an entirely alien way of thinking. Consider the suffering involved in childbirth. I've heard from some women who have had children that the torment is an important aspect of it, that the relief provided by an epidural would make the experience less meaningful, less authentic.

THE EFFORTS BY religion to explain unchosen suffering often resonate. They are what we want to hear, and they mesh well with psychological systems that are seeking to find meaning in the worst of events. But in other regards, such explanations are hard to accept. The psychological immune system has its limits. Many of the features that make suffering so rewarding when it's chosen—what we've been talking about so far in this book—are absent when it is involuntary.

One pleasure of chosen suffering is that it can be a form of play. But there is no such thing as unchosen or unwanted play; being forced to play isn't play. This point is illustrated in an anecdote reported in a psychiatric journal, one that takes us back to dentistry. There was a woman who had a high desire for pain in S&M sessions with her boyfriend but who hated going to the dentist. Her boyfriend tried to get her to construe a dental exam as yet another erotic masochistic adventure, but she would have none of it—for her, there was no getting around the fact that visiting the dentist was something she had to do; it was not a free choice. She couldn't tell a story, after the fact, that would transform this into chosen suffering.

Or consider the pleasure of mastery. This is naturally found in chosen suffering. C. S. Lewis makes this point with regard to fasting in *The Problem of Pain*, albeit disapprovingly: "Everyone knows that fasting is a different experience from missing your dinner by accident or through poverty. Fasting asserts the will against the appetite—the reward being self-mastery and the danger pride." But there is no self-mastery when you don't choose to be hungry.

And then there's morality. Religions often claim that it is morally good to suffer. This might make sense for chosen suffering, where the suffering is a voluntary act. But you don't normally get moral credit for acts you don't choose. If I give most of my money to the poor, this is a sacrifice that I can take satisfaction in. But if the poor take my money against my will, it's hard to tell a story in which I am a hero.

Still, one might show goodness in how one responds to the unchosen suffering. It can be moral to be stoic and brave in the face of suffering, not to complain too much, not to try to move your burden onto the shoulders of others. In other circumstances, the expression of suffering can show off a moral sensibility. An anonymous British pamphlet from 1755, *Man: A Paper for Ennobling the Species*, proposed a number of ideas for human improvement, and among them was the idea that something called "moral weeping" would help: "Physical crying, while there are no real corresponding ideas in the mind, nor any genuine sentimental feeling of the heart to produce it, depends upon the mechanism of the body: but moral weeping proceeds from, and is always attended with, such real sentiments of the mind, and feeling of the heart, as do honour to human nature; which false crying always debases."

Sometimes tears are the right response. I am close to a man who, after losing his wife to cancer, became depressed in the months that followed. Over time, several people suggested that he seek some sort of help—a therapist, maybe, someone who could get him some antidepressant medication. Not surprisingly for anyone who knows him, he refused. He could see that happening in the future, maybe, but he felt his sadness was right, it was properly respectful, and it would be wrong to make it go away. I'm not sure I have quite the same attitude, but of course I would hardly want to be out dancing if someone close to me had just died. It's not just that it would look bad to others; it would be morally grotesque.

In her discussion of having children and the savage risk it involves, Zadie Smith quotes Julian Barnes, who told her about a condolence letter he once received that said, "It hurts just as much as it is worth." Pain can be a proper acknowledgment of value.

Some of our thoughts about the relationship between suffering and goodness are not rational. In 1994, Daniel Pallotta founded the company Pallotta TeamWorks to do fundraising for causes such as research on AIDS and breast cancer, and he was successful, bringing in more than $300 million in nine years. But the company was not itself a charity; Pallotta made money from his fundraising, about $400,000 a year. When this was reported, people were appalled. Organizations were pressured to stop working with him, and ultimately his company closed.

This event fascinated two colleagues of mine, George Newman and Daylian Cain, and they decided to investigate "tainted altruism"—the discounting of altruistic acts that give us personal gain, even if they make the world better. In one of their studies, people read about a man who, to gain the affection of a woman, spent several hours a week volunteering where she worked. Some subjects were told that this was a homeless shelter; and it was emphasized that, though the man was self-interested, he did a good job at helping out. Others were told that it was a coffee shop. Subjects judged him to be a worse person when he worked at the homeless shelter. Closer to the Pallotta case, they also found that subjects judged someone more harshly for running a charity for profit than for running a corporation for profit.

We want our goodness to be unsullied by pleasure. Now, this isn't quite the same as showing that we connect good works with suffering—but there is evidence for this stronger claim as well. Think about the Ice Bucket Challenge, a viral social media campaign that encouraged people to pour a pail of ice water over their head to support ALS research. In experimental work, people tend to contribute more to a charity when they expect to endure pain and suffering for that cause—the so-called martyrdom effect. The willful suffering we see in religion—fasting, sacrifice, even self-mutilation—might well reflect a more general feature of what we see as virtuous. It isn't good if it doesn't hurt, so when we do good, we are willing—in fact, eager—to experience pain. This is why savvy charities sponsor walkathons and marathons, not group massages and beach parties.

I will point out a twist here. It's not enough to simply suffer; the pain has to have meaning. Christopher Olivola and Eldar Shafir provide an example that makes that point. Suppose you have a friend who is sick and too weak to clean her house. During a visit, she is resting in another room, and you decide to surprise her by washing her dishes, which are piled high in her sink. You work an hour on this, scraping, washing, rinsing, and drying. Just as you finish the last dish, your friend enters the kitchen and discovers the clean sink. This seems like a nice moment. Look what you've done for her! Even if you left and she never realized that it was you who did all of this, you might still feel some sort of warm glow.

But imagine that your friend now reveals that the kitchen has a brand-new dishwasher, which, if you had seen it, would have enabled you to quickly do all the dishes, getting the same result with less effort and time. You would feel, Olivola and Shafir suggest, less satisfied.

They find support for this idea experimentally. If you ask one group of people whether they will participate in a charity that involves a five-mile run (grueling) and a second group whether they will participate if the event involves a picnic (pleasant), the people in the first group are more likely to agree. This is the martyrdom effect. But when people are asked to consider both options at the same time, to choose between the five-mile run and the picnic, they tend to choose the picnic. Presumably they reason that since the picnic will do just as much good, there's no point to the extra suffering of the five-mile run. As the authors point out, this suggests that we don't simply have a "taste for painful benevolence"; rather, suffering is valuable, but only when it's seen as essential for a positive result.

—

We've been talking about whether we think unchosen suffering is good for us, whether it improves our relationship with God, teaches us valuable lessons, allows for spiritual growth, and is morally good. But forget about what people believe—what about truth? Is unchosen suffering actually good for us? Does it really make us more resilient, nicer, better people?

Many think it does. Making the case, here's part of a much quoted commencement speech by Chief Justice John Roberts, from 2017.

> Now, the commencement speakers will typically also wish you good luck and extend good wishes to you. I will not do that, and I'll tell you why. From time to time in the years to come, I hope you will be treated unfairly, so that you will come to know the value of justice. I hope that you will suffer betrayal because that will teach you the importance of loyalty. Sorry to say, but I hope you will be lonely from time to time so that you don't take friends for granted. I wish you bad luck, again, from time to time so that you will be conscious of the role of chance in life and understand that your success is not completely deserved and that the failure of others is not completely deserved either.

For Roberts, suffering gives perspective and nurtures empathy. A related view is that it builds resilience—it makes one, in Nassim Nicholas Taleb's nice phrase, "anti-fragile." This is expressed in the famous aphorism of Nietzsche: "That which does not kill us makes us stronger." Brock Bastian put it in more formal terms: "The key to healthy psychological functioning is exposure."

There is research that favors this view. One study, by Mark Seery and his colleagues, went like this: Subjects were first given a list of thirty-seven negative life events—physical assault, death of a loved one, and so on—and tallied up how many they had experienced in their lives. Then subjects put one of their hands in ice-cold water and answered questions about how intense the pain was, whether it worsened their mood, and whether they were prone to "catastrophizing" (which meant that they would agree with statements like "I thought that the pain might overwhelm me"). The experimenters also measured how long they left their hands in the water.

People reported between zero and nineteen lifetime adverse events, with 7.5 percent reporting none at all. Lucky devils? Well, maybe not. The data showed an inverted U-shaped curve—the people who dealt best with the pain were those who had a midrange of exposure. The people who had made it through life scot-free were, relatively speaking, wimps.

They did a second study with a different method, where the stressor wasn't immersion in cold water. It was navigating a computer obstacle course that was described to them as an important test of nonverbal intelligence. And stress wasn't measured with questionnaires; it was done through a series of physiological measures, including heart rate. The results were identical: the most positive reactions weren't from those with no stress in their life or with lots of stress in their lives; they were from those who were in between, in the sweet spot.

A similar study was done by David DeSteno and his colleagues. This one focused on kindness. As in the previous study, they asked people about how much adversity they had faced in life. They also measured what's called "dispositional compassion," using a standard scale with five items assessing how much people agreed with claims like "It's important to take care of people who are vulnerable" and "When I see someone hurt or in need, I feel a powerful urge to take care of them." Finally, they included a chance for the subjects to donate actual money. It turns out that experiences of past adversity are associated both with the expression of compassion and with actual money given. This meshes with other work suggesting that poor people, who tend to experience more stress and adversity, show high levels of compassion on a variety of measures.

We should be cautious here. These effects are statistically real, but subtle. And cause and effect are hard to pull apart. Perhaps there is some third factor that influences both your propensity to experience certain negative life events and your resilience and kindness.

Still, there is at least anecdotal evidence for the social benefits of suffering, at least in the short term. Rebecca Solnit, in her book *A Paradise Built in Hell*, chronicles how groups of people respond to disasters, arguing that they are far kinder to one another than you would expect if you read Hobbes, who maintained that, stripped of external constraints, people would descend into savagery. Actually, Solnit says, you find that "the prevalent human nature in disaster is resilient, resourceful, generous, empathic, and brave." For her, disaster provides an opportunity. People don't just rise to the occasion; they do so with joy. This reveals "an ordinarily unmet yearning for community, purposefulness, and meaningful work that disaster often provides."

You can see some sign of this in the laboratory. Brock Bastian and his colleagues did a series of experiments where some of the participants put their hands in ice water, performed leg squats, and ate hot chili peppers. They did this in small groups, and those members of groups who shared a painful experience felt more bonded, trusted one another more, and cooperated more with each other.

SOME AMOUNT OF suffering in life might be a good thing, increasing resilience and kindness and bringing people together. But what about specific terrible events, such as being raped or losing a child to cancer? Can they have actual positive effects? Do we benefit from unchosen suffering, as we benefit from chosen suffering?

It would be perverse to insist that bad events can never have good outcomes. James Costello says that his life was much improved by being maimed in the Boston Marathon bombing, and who am I to say that he is wrong? Yes, his view is probably shaped by a psychological bias to find benefit in bad events, but still, this doesn't mean that he's mistaken. He did meet the woman of his dreams.

There is a lot of unpredictability in the world, after all. Imagine you get a severe flu and have to miss flying to London for your best friend's wedding. You might think back on this as a great loss—but maybe if you had gone, you would have wobbled out of your hotel the next morning, happy and hungover, started to cross the street, looked left instead of right, and gotten hit by a double-decker bus, the ones with all the tourists on top. The flu might have saved your life. Here's a Taoist story you may have heard:

> [There was] an old farmer who had worked on his crops for many years. One day his horse ran away. Upon hearing the news, his neighbors came to visit. "Such bad

luck," they said sympathetically. "May be," the farmer replied.

The next morning the horse returned, bringing with it three other wild horses. "How wonderful," the neighbors exclaimed. "May be," replied the old man.

The following day, his son tried to ride one of the untamed horses, was thrown, and broke his leg. The neighbors again came to offer their sympathy on his misfortune. "May be," answered the farmer.

The day after, military officials came to the village to draft young men into the army. Seeing that the son's leg was broken, they passed him by. The neighbors congratulated the farmer on how well things had turned out. "May be," said the farmer.

But those who talk about the transformative power of negative events aren't merely saying that the world is unpredictable and that seemingly bad events can have positive outcomes. They want to make the stronger claim that we are constituted in such a way that, for some of us at least, some bad things are actually good for us.

I think we should keep an open mind here. But I also think that, as David Hume put it, "extraordinary claims require extraordinary evidence." And this is truly an extraordinary claim.

Often when a criminal is sentenced to death, advocates against the death penalty will talk about the awful life that this person has had; they will tell stories of horrible abuse in childhood and cruel treatment as an adult. They will try to make the case that these experiences have damaged the defendant somehow, and we should accordingly show mercy. Whether or not this argument works depends on what you think about the death penalty, moral responsibility, and forgiveness, but the form of the argument makes sense: people realize that a horrible life can mess one up. Nobody responds to these stories and says, "Well, then the punishment should be *worse*, because all these bad experiences should have made the person kinder and more resilient than the rest of us!" This would really be a perverse response.

Bad experiences wound us; they make us bitter and fearful, more self-protective and less kind. There is trauma and PTSD. Of course, people can be resilient and, when enough time goes by, might be emotionally healthy after awful life events. But still, these events *are* negative. I'm trying to be good about supplying references as I write this book, providing scientific citations in the endnotes that support the factual claims I make, but really, it's strange to need to provide empirical support for the view that being raped or tortured is bad for psychological functioning. It's like needing support for the claim that being hit by a car is bad for the body.

THERE ARE THOSE who would push back at the skeptical position I just laid out. For instance, some scholars are intrigued by the possibility of what's been called "altruism born of suffering." The idea is that people who have suffered—and much of the work here has focused on suffering at the hands of others, as in neglect, physical assault, sexual abuse, and torture—often become motivated to help others, not in spite of their abuse but because of it.

People do often describe their kindness this way. In a recent review article, Johanna Vollhardt gives many examples of this. Victims of spinal cord injury or stroke describe how they were motivated to help others with similar afflictions. Rape victims would work for organizations that helped other rape victims. Mothers Against Drunk Driving was founded by a mother who lost her child in an accident caused by a drunk driver. And so on.

Vollhardt offers some explanations for how suffering might have this effect. Helping others might distract sufferers from their own problems and might even cheer them up. Helping others might put our own suffering in a different light; when you interact with people as bad off as yourself or even worse, your own problems might seem smaller. It might make you feel more competent and efficient. It might better integrate you into a social community. Most of all, building on the ideas of Viktor Frankl, helping others might give your own suffering more meaning and value; it might infuse it with purpose. There is a big difference between *This terrible thing happened, and I suffered and that's it* and *This terrible thing happened, and I suffered, but then, because of that, I ended up helping others and made a positive difference in the world.*

Theoretically, all of this makes sense. But there is little actual evidence that sufferers are kinder than they would have been had they not suffered. Many of the studies that purport to show this have small samples; some are just case studies of single individuals. Often they rely on self-reporting, not on any sort of objective measure. For instance, one study interviewed one hundred Holocaust survivors, and most of them described how they had helped others, providing food and clothing to others in the camps. But what are we to make of such stories? Even if we weren't naturally prone to overstate the amount of good that we do, memories of events in the distant past are inaccurate, and we often remember our past actions as being far more positive than they actually were.

Most of all, these studies rarely have control groups. You learn, say, that many people who have experienced the death of their child end up working for a good cause later on. But what would the same people have done if they hadn't experienced such a tragedy? And how many people who suffered the death of a child have withdrawn from the world and become cold and unkind as a result, or were so damaged by the process that they were unable to be a positive force in other people's lives?

WHAT ABOUT RESILIENCE and toughness, then? In laboratory studies, we've seen hints that some degree of suffering in one's life makes one more resilient. What about in the real world?

This was the topic of a study by Anthony Mancini and his colleagues. It was a "natural experiment" in which they took advantage of a tragic event to explore the effects of trauma. I'm going to spend a bit of time describing this study, because it's important and well known and interesting, and also because it illustrates some problems that arise when asking this sort of question.

In 2007, at the Virginia Polytechnic Institute, a mentally ill student went into the student residence and murdered thirty-two people and wounded twenty-five others. The massacre lasted over two hours. It turned out that there was an ongoing study on women on the campus (having to do with sexual victimization), so Mancini and his colleagues had access to all sorts of psychological measures before the shooting and were in an ideal position to do a before-and-after comparison. They already had data from 368 women and collected more data two months after the shooting, and then again after six months and one year.

Looking at depression scores, the women fell into four main classes: 56 percent were fine—not depressed before the shooting, not depressed after. Eight percent were the opposite—miserable before and miserable after—while 23 percent got worse over time: they were fine prior to the shooting, but in post-test they were depressed. And—this is what most impressed the authors of the paper—13 percent got better. Depressed before, less depressed afterward. The ratings for anxiety were similar, though a smaller proportion improved: 7 percent.

What explains the cases where people got better? Mancini and his colleagues suggest that it resulted from the social support that came after the mass shooting. All of this love and concern and therapy ended up helping students who were depressed and anxious—not just with any trauma that might have occurred as a result of the event but, as a lucky accident, with other life problems that the students had as well. Mancini and his colleagues point out that this is a feature of events like mass shootings: "A key aspect of a mass trauma is that it afflicts large numbers of people at once and therefore can mobilize mutually supportive and cooperative behaviors on a broad scale." This is different from individual traumas, such as rape or assault, that don't mobilize communities and might actually leave the victims feeling isolated and alienated.

All of this makes sense. But there are a few concerns to keep in mind.

First, the data they report likely overstate the benefits of the traumatic event. Many people contacted after the event didn't complete the follow-up surveys, even though they were offered compensation, such as gift cards, charitable donations in their name, and being entered into a lottery with a chance to win money. People who are traumatized by an event are going to be less likely to do the follow-up surveys, so the results over-represent those who had the most positive outcomes.

When we forget about those who drop out, we risk succumbing to survivorship bias. The best illustration of this bias is the story about the American military in World War II. They wanted to add armor to the planes to protect the pilots, but they also had to minimize the weight of the planes, so they needed to add the armor only where it would do the most good. The officers in charge inspected the planes that returned from air battles, looked at where the bullet holes were, figured that these were the areas where the planes tended to get hit the most, and recommended that the armor be placed there. When I read the story and got to this part, I thought it made perfect sense.

Which means that I'm nowhere as smart as the statistician Abraham Wald, who told them where their logic was wrong. The planes they were studying were those that had returned. This means that being riddled with bullet holes in certain places is actually not a bad thing. It's *the other locations* that you should put armor on, because apparently if you get hit in some of these places, you don't come home.

The logic here is more general. Suppose you have a psychology graduate program where many students drop out. You look at the students in their final year who are struggling the most and discover that they are, say, poor in statistical skills, and decide that it's a top priority to improve these skills. This is a mistake. It is putting armor in the areas where there are bullet holes in the planes that return. You should actually make the *opposite* inference—apparently, you can make it to the end of the graduate program with bad statistical skills, so this can't be what's most important.

This issue is particularly pressing when we explore the effects of terrible events by studying people who have experienced such events and see how well they are doing. When we do this, we are typically excluding people in hospitals, psychiatric institutions, and prisons. We are certainly excluding people who are unwilling or unable to participate in psychological studies. All of this is inevitable, and there's nothing inherently wrong with it—so long as we don't forget about the survivorship bias and are cognizant that, by excluding those who are the most damaged, we will exaggerate any positive effects of the experience.

Also, in the Virginia Polytechnic Institute study, there was no control group—no comparison group of subjects for whom there was no mass shooting. This is understandable under the circumstances, but it means that we have no idea what would have happened to these women if the shooting had never happened. Nobody is surprised that some number of young people who go to college become depressed; what's so stunning about the idea that some number of young people become less depressed? Is it really so crazy to think that about one in eight college students might show improvement regardless of the shootings?

And suppose this is wrong; suppose that this small number of students wouldn't have improved if not for the mass trauma—and let's forget also about the survivorship bias. Still, under Mancini and his colleagues' own theory, it wasn't the event itself that caused any improvement. It was the mobilization of social services, all the love and attention, that was associated with the event. Suppose you are in a mild car accident, go to physiotherapy, and, as a result of getting some treatment, start to exercise more, eat better, and take better care of yourself. This doesn't mean that car accidents are good for you.

Taking the results at face value, then, it suggests that support is good for depressed and anxious students. The fact that such support was made available because of a mass shooting is just a perverse twist of fate.

Some researchers are interested in post-traumatic growth—the idea of general positive changes as the result of terrible events. This is different from the psychological improvements we just discussed, where the focus was recovery from depression and anxiety. And it's different from resilience, which means being unscathed. Post-traumatic growth is *improvement.* As Richard Tedeschi, one of the founders of the theory, puts it, "People develop new understandings of themselves, the world they live in, how to relate to other people, the kind of future they might have and a better understanding of how to live life."

A commonly used scale looks for improvement in five areas:

1. Appreciation of life
2. Relationships with others
3. New possibilities in life
4. Personal strength
5. Spiritual change

There are many anecdotes about people who claim to have experienced post-traumatic growth. Perhaps you have a story yourself about how some awful experience led you to better appreciate life, improve your relationship with others, find God, and so on. One might be skeptical about any given case—if we are prone to see the bright side of things, then these stories will come naturally to us, regardless of whether they are true. Still, it would be foolish to doubt that, sometimes, trauma can bring out positive personal transformation.

On the other hand, one can doubt that there is a general process of post-traumatic growth. Consider the findings of a recent meta-analysis called "Does Growth Require Suffering? A Systematic Review and Meta-Analysis on Genuine Posttraumatic and Postecstatic Growth." There are three main conclusions from this review of the literature:

1. There is some evidence from prospective studies—studies that collect data before and after the traumatic event—that there is some improvement, after a traumatic event, in self-esteem, positive relationships, and mastery. There is no growth in the categories of meaning and spirituality.
2. *But* these effects are just as powerful after major positive life events as after major negative events.
3. *And* they probably have nothing to do with the events themselves. Many studies don't have control groups; they don't compare what happens after the positive or negative experience with what happens if there is no event at all. When the authors of the review looked at the studies that had control groups, they found that most show no effect. That is, people tend to say that they got better in some regard after a major life experience, but they also say that they got better during the same time period if there was no experience at all.

Again, nobody is doubting that terrible events can lead to positive personal changes. But it also turns out that wonderful events can lead to positive personal changes, and that, perhaps just as often, the absence of any event at all is also followed by positive personal changes.

I've talked throughout this book about the importance of chosen suffering, its role in pleasure and meaning, and I'll say more about this in the next and final chapter. But my assessment here of unchosen suffering is less positive. We try to tell stories about its value, and some of these stories may have some truth—we've seen some evidence that some amount of suffering in your life does make you kinder and more resilient. And it might be psychologically useful to try to find benefit in loss and pain. Still, this is a case where common sense is right: we are smart to try to avoid cancer, mass shootings, the death of our children, and other horrors.

After all, even if suffering does have its benefits, it's pretty likely that enough of it will come to you and those you love regardless of what you do. You don't have to look for more.

7

SWEET POISON

> I have two luxuries to brood over in my walks, your Loveliness and the hour of my death. O that I could have possession of them both in the same minute. I hate the world: it batters too much the wings of my self-will, and would I could take a sweet poison from your lips to send me out of it. From no others would I take it.
>
> —JOHN KEATS, LETTER TO FANNY BRAWNE, JULY 1819

We are going to die, and that makes us the lucky ones," writes Richard Dawkins. After all, we're the ones who got to exist in the first place.

For this we have our ancestors to thank. Each of them, for the past 3.8 billion years, "has been attractive enough to find a mate, healthy enough to reproduce, and sufficiently blessed by fate and circumstances to live long enough to do so." To make it this far, we have to be pretty hot stuff, though we should be humbled that the same victory lap could be taken by every creature we share the world with, every rat, goldfish, and mosquito—all survivors of a multibillion-years-long game of battle royal.

Humans are built for success in ways that are both shared and unique. Like many other creatures, we are constituted to know things about the world. Animals that believe true things do better, on the whole, than those who don't. If there is a cliff to your right, if your tribe is getting sick of you, if there's something biting your leg, it's good to know about it. This is what eyes and ears and other sensory organs are for, along with big chunks of our brains. Your primate competitors in the past who weren't as good at forming true beliefs didn't make it to the next round.

Unlike other animals, though, humans also possess a moral sense. All normal humans possess some capacity for kindness and a sense of fairness and justice—along with darker moral facets such as resentment, outrage, and an appetite for vengeance. This, too, has a logic to it, enabling large groups of unrelated individuals to constrain their nasty and destructive impulses and work together for mutual gain.

We are imperfect beings, though: "fallen angels, not risen apes," in the words of the anthropologist Robert Ardrey. Evolution hasn't built us to learn true things as a goal in itself, but to serve the goals of survival and reproduction. And so we have no natural access to truths about the distant past and the distant future, or the very small (such as subatomic particles) and the very large (such as galaxies). We are unprepared to cope with metaphysical questions about free will, causality, or the nature of consciousness. Such knowledge is, from the standpoint of our genes, useless. We are also subject to bias. When truth and utility clash, truth comes in second, which is why we often have irrational fears. Ask anyone who is afraid of spiders and snakes.

We are similarly limited in the realm of morality. We possess a Flintstones morality. We aren't built to appreciate the immorality of racism, or to understand that, from an objective point of view, the happiness of unfamiliar children thousands of miles away is of the same importance as the happiness of our own children. Just like knowledge about subatomic particles, an impartial morality is not what our minds have evolved for—it has no adaptive value.

But somehow humans—and only humans—have done something astonishing. We can transcend our limitations. We have developed science, technology, philosophy, literature, art, and law. We have come up with the Universal Declaration of Human Rights; we've been to the moon. We use contraception, deliberately subverting nature's goal of reproductive success so that we can pursue other goals. We give some of our resources (nowhere near enough, but some) to strangers, overcoming our biological drive to favor family and friends.

We don't marvel at this enough. It's so odd that this could have ever happened, that minds that evolved to cope with a world of middle-size objects—plants and birds and rocks and things—could come to have some grasp of the origins of the universe, quantum forces, and the nature of time; that minds that evolved to feel kindly toward kin and to be grateful to those who treat us kindly could arrive at moral precepts that motivate charity for those far away.

Some people think that all of this is a miracle, actually, and therefore proof of the existence of a loving God. I am skeptical myself, and in other work I've argued directly against this theistic analysis in the domain of morality. But it's a tempting inference to make. I'm the least spiritual person you're likely to meet, but still, if I wanted to talk myself into belief in divine intervention, looking at this sort of transcendence is where I would start.

WE'VE TALKED ABOUT truth and goodness, but what about pleasure and meaning? How do these capacities fit into this picture of our evolved natures?

I think one can tell a similar story. Our moods and feelings, our rhythm and blues—these, too, are the products of brains evolved through natural selection. We feel relief when things go well, afraid when threatened, filled with grief at the death of a loved one. Such emotions are adaptations that increase our odds of survival and reproduction. Thank (or blame) Darwin for all this.

The details of how our emotions and feelings serve adaptive purposes can be found in countless evolutionary psychology analyses, most of which focus on short-term pleasures linked with reproductively relevant goals, such as nourishment and status and procreation. But long-term moods like happiness can be seen in the same way. As Steven Pinker writes,

> We are happier when we are healthy, well-fed, comfortable, safe, prosperous, knowledgeable, respected, non-celibate, and loved. Compared to their opposites, these objects of striving are conducive to reproduction. The function of happiness would be to mobilize the mind to seek the keys to Darwinian fitness. When we are unhappy, we work for the things that make us happy; when we are happy, we keep the status quo.

A hard truth arises from this. We are not built to be happy. Evolution doesn't want us to be in constant bliss any more than it wants us to be pain-free. Pain is information about what's wrong and an inducement to make things better. Sadness and loneliness and shame play similar roles.

But not all of our negative feelings are useful. It would be a blessing to shut off chronic pain when there's nothing that can be done about it, or to cure disorders of depression and anxiety. And sometimes our negative feelings are poorly calibrated to our lives right now. As Robert Wright notes,

> Modern life is full of emotional reactions that make little sense except in light of the environment in which our species evolved. You may be haunted for hours by some embarrassing thing you did on a public bus or an airplane, even though you'll never again see the people who witnessed it and their opinions of you therefore have no consequence. Why would natural selection design organisms to feel discomfort that seems so pointless? Maybe because in the environment of our ancestors it wouldn't have been pointless; in a hunter-gatherer society, you're pretty much *always* performing in front of people you'll see again and whose opinions therefore matter.

A similar way in which evolution has gone awry involves what is sometimes described as the "hedonic treadmill." Increases in pleasure are short-term; you are delighted by a new experience or event, but then, over time, you go back to where you were before. The first kiss is great, the thousandth less so. No matter how fast you run, you are still in the same place.

This is often seen as a version of a more general psychological truth discussed at the very start of the book—our minds respond to change; we habituate and grow insensitive to the status quo. But there's likely to be something more specific at work. A creature that could savor positive experiences indefinitely might stop striving, and hence be at a disadvantage relative to those who are less prone to stand pat. Some degree of unsettledness, anxiety, and ambition may be baked into the human condition. And much of this is connected to status—where you stand relative to others. I'm happy with my car, but then my neighbor gets a nicer one and my happiness goes away.

In such cases, evolution's goals (metaphorically speaking, as always) are not our own. They are not the goals and priorities that we, as contemplative beings, should aspire to. I don't want to have as many children as possible. I'd rather not care so much about what strangers think of me. And I'd certainly prefer not to be forever dissatisfied with my life.

Fortunately, we are not stuck with our initial settings. We can game the system. Just as we can recognize that our eyes are limited in their powers and build telescopes, just as we can worry that our morality is biased and so work to establish impartial procedures of justice, similarly, we can also grow frustrated with the carrot-and-stick nature of our feelings and try to do better.

Is there something wrong about thumbing our noses in the face of natural selection? Shouldn't we want to do what we have evolved to want to do? Shouldn't we feel as we have evolved to feel?

No. This view is a fallacy. There is no logical connection between *This is how things are* and *This is how things should be*. After all, such an inference leads to absurd conclusions. It implies that a man whose sole activity is donating to a sperm bank is living an infinitely better life than, say, the Dalai Lama, who is childless. Or that a woman who gives birth to many children and treats them terribly (but they do survive and reproduce) is living a much better life than if she adopted children and treated them with love and respect. I can't imagine dumber positions to take.

Now, *some* attempts to hack one's mind are foolish and immoral. We have evolved to get pleasure from social contact, productive activities, meaningful relationships, and so on, but it is possible to shortcut this with drugs and alcohol. Maybe in the future, people will have lives of great pleasure that are spent in a blissed-out state, without any of the bad effects that drugs like heroin have. But this is the wrong goal; it's a wasted life. Similarly, if there were a psychopath pill, some people would probably take it, pleased to be liberated from their consciences, despite the damage this would do to other people. And a lot of us would be tempted by a pill that would strip away anxiety and sadness, even if, in the long run, this, too, led to a less full life.

But other approaches are more promising. I quoted Robert Wright above on how we're evolved to react in disproportionate ways to what strangers feel toward us. He suggests that Buddhist meditative practices can be a fix for this. More generally, he sees in Buddhism a rebellion against the priorities that natural selection has wired into us. Because of evolution, we are driven by attachments and passions; we worry, obsess, and plan. Our perception of the world is colored and clouded by our desires. But meditation might cure all this. We might come to appreciate the world as it is, to banish the ego, to rid ourselves of unhealthy attachments.

There is much enthusiasm for this approach these days, in both academia and popular culture, and I agree that it deserves more investigation, including more empirical studies on the effects of meditation. But since nobody else seems to be arguing against it, I'll add a critical remark.

My concern is about our relationships with friends and family. The Buddhist goals of equanimity and nonattachment have great moral appeal. My last book was called *Against Empathy*, and I argued there that emotions such as empathy are too biased and innumerate and parochial to be good moral guides; we are better off, when making important decisions, with a more distanced approach, what I called "rational compassion." And I drew upon Buddhist ideas to make this argument; in this regard, my book was quite aligned with Buddhist philosophy.

But in the book, I also wrestled with challenges to my view, and one of these was the issue of close relationships. Rational compassion seems antithetical to being a loving parent, friend, or romantic partner. You're not supposed to be distant and unbiased toward those you love. Being a good father, for instance, involves prioritizing one's children over other people's children; it means caring about them and loving them more. To the extent that Buddhism denies the specialness of close relationships, it's missing something important. (Recall the old joke: "Did you hear about the Buddhist vacuum cleaner? It comes with no attachments.") I'm not doubting (of course) that Buddhists, or non-Buddhist practitioners of Buddhism, can be caring parents, friends, and lovers. But to the extent that they are, they are falling short of the precepts of their discipline.

For better or worse, my goal in this book has been more modest. I am not suggesting that we transcend our natures. Rather, I've been considering what gives us pleasure and happiness and fulfillment, and looking at where suffering fits into all of this. This has been done in the spirit of exploration and curiosity, trying to make sense of certain aspects of our natures. But still, there are practical implications that follow from this exploration, some suggestions for how to best live our lives.

One implication brings us back to motivational pluralism. There is a classic contrast between pleasure and meaning, hedonia and eudaemonia. Which should we choose? It turns out that one can have both. Actually, the conclusion is even stronger—it's not merely that there exist some people who are both happy and have lives with meaning. It's that there is a correlation: happy people are more likely to say that their lives are meaningful, and people who say that their lives are meaningful are more likely to say that they're happy.

One series of studies looking at the relationship between pleasure and meaning asked college students to integrate either new pleasurable experiences or new meaningful experiences into their lives. The pleasurable activities included extra sleep, shopping, watching a movie, and eating sweets. The meaningful activities included helping another person, taking time to introspect, and striving to have a meaningful discussion with someone.

The researchers found that both of these additions had positive effects. Adding pleasure to a life that students reported was already full of meaning gave them increased good feelings and carefreeness, while adding meaningful activities to a life full of pleasure led to "greater elevating experience." And when the students were asked to include both types of new activities, there were multiple benefits. As the authors summarize, "Given that we expected both eudaemonia [meaning] and hedonia [pleasure] to contribute to well-being in life, and that we did not see them as mutually exclusive, we expected their combination to be linked with particularly great well-being. We found good support for this prediction. People who pursued both eudaemonia and hedonia reported higher levels of most well-being variables than people with neither pursuit."

We should be cautious here. These effects are not strong, and this is a college population, not a broad sample of humanity. Still, it fits well with what we know from other studies. When it comes to happiness and meaning, the tagline from the old light beer commercial rings true: You can have it all.

BUT YOU CAN also mess things up. If motivational pluralism is true, and I think it is, then a too narrow focus on one sort of motivation can have bad effects.

In particular, it turns out that one can screw up being happy by *trying* to be happy—or at least by trying to be happy in the wrong way. There are studies that look at the extent to which people are motivated to pursue happiness, by asking them to rate themselves on items such as "Feeling happy is extremely important to me" and "How happy I am at any given moment says a lot about how worthwhile my life is." The people who highly agree with such items are less likely to get good outcomes in life and more likely to be depressed and lonely.

Now, the usual worries about direction of causation apply here. Maybe it's not that trying to be happy makes people depressed and lonely—maybe it's that depressed and lonely people are more motivated to try to be happy. But there is some experimental work that supports the corrosive attempts of seeking out happiness. In one study, people were asked to make themselves feel happy while listening to Stravinsky's *Rite of Spring*. Compared with people who simply listened to the music, their mood got worse. Another study found that after reading an article discussing the advantages of happiness—which presumably makes people value happiness more—subjects were less happy after watching an enjoyable film clip. Focusing on happiness does seem to have a bad effect.

The psychologists Brett Ford and Iris Mauss have some suggestions as to why this is so. Maybe when you pursue happiness you set unrealistically high standards for success, setting yourself up for failure. Or maybe the self-conscious pursuit of happiness makes you think a lot about how happy you are, and this gets in the way of being happy, in the same way that thinking about how good you are at kissing probably gets in the way of being good at kissing.

The most plausible explanation, and the one they stress the most, is that people aren't accurate about what makes them happy. It turns out that pursuing extrinsic goals related to praise and reward—looking attractive, making money, and building up social status—makes you less happy and less fulfilled, and is linked with more depression, anxiety, and mental illness. One meta-analysis, summing up more than 258 studies, found that "respondents report less happiness and life satisfaction, lower levels of vitality and self-actualization, and more depression, anxiety, and general psychopathology to the extent that they believe that the acquisition of money and possessions is important and key to happiness and success in life."

(Yes, I know I said earlier that money is related to happiness. There is no contradiction here. Money does make you happy; it's the *trying* to make money that makes you sad. The trick is to get money in the course of other, meaningful, pursuits—or, if you can manage it, to be born into wealth.)

The problem, then, might not be trying to be happy, but rather trying to be happy in certain ways. And indeed, cross-cultural research finds that in collectivist societies, such as parts of East Asia, trying to be happy *is* related to happiness, presumably because these efforts are much more socially engaged and connected with friends and family. It is in more individualistic societies like the United States, which have more of a materialistic ideology, that pursuit of happiness brings you down, because we go about it in the wrong way.

What about becoming a hedonist? The duration of felt experience—our feeling of *right now*—is between two and three seconds, about how long it takes Paul McCartney to sing the words "Hey Jude." Everything before this is memory; everything after is anticipation. So what about a life dedicated entirely to improving this moving window of two to three seconds? To put it in the language we introduced earlier, this would be a life devoted to nothing but experienced happiness. I argued in the first chapter of this book that we are not natural hedonists, that we have multiple goals that we aspire to. But maybe we should be hedonists! Maybe our lives would be better if we focused more on pleasure.

I think this would be a mistake as well, a second way to

screw things up. But this view has some sharp defenders. My favorite argument comes from Dan Gilbert, who starts off with an example:

> So I may be a shameless hedonist happily swimming in my Olympic size pool, feeling the cool water and the warm sunshine on my skin and my hedonic state could only be described as pleasurable. Occasionally I jump out of the pool, pause, and think about how empty my life is, and for a few minutes I feel bad. Then I get back in the pool and swim some more.

If we continue to spend our days in the pool, we will have a life filled with experienced happiness but without overall satisfaction and without meaning. How bad would this be?

As we've seen earlier in this book, many people think that this wouldn't be a good life. Describing his career to Tyler Cowen, Daniel Kahneman says, "I was very interested in maximizing experience, but this doesn't seem to be what people want to do. They actually want to maximize their satisfaction with themselves and with their lives." Dylan Matthews says something similar: "I think it's fair to say that this metric—life satisfaction—is a better gauge for what people actually want for themselves than emotional well-being is. I don't want to be perpetually giddy and worry-free; I do want to have a life that I'm, on the whole, happy with." And, then, of course, there is John Stuart Mill: "It is better to be a human being dissatisfied than a pig satisfied; better to be Socrates dissatisfied than a fool satisfied. And if the fool, or the pig, is of a different opinion, it is only because they only know their own side of the question."

Gilbert is unpersuaded by these reactions. He points out

that in his pool example, there are two different sorts of conscious experiences, which we can see as akin to two different people. There is the Experiencer, who feels the cool water and the warm sunshine and who is happy. And there is the Observer, who passes judgment on the life as a whole and who is disappointed.

Gilbert notes that when we consciously mull over the question, we call into existence the Observer, the part of ourselves that's the inner Socrates. This might give the impression that we are always Observers. But, as Gilbert puts it, this would be like someone who concludes that the refrigerator light must be on all the time because it's lit up whenever they open the door. Indeed, Gilbert notes that the Observer is rarely present in our lives. We spend little time thinking of our lives as a whole. When you are in the pool, with the cool water and warm sunshine, or laughing with friends (or, for that matter, undergoing a painful dental procedure or falling down a flight of stairs), you aren't evaluating your life. You are living it—you are the Experiencer.

So we consult the Observer and hear, "This is not a valuable life. I'm very disappointed." But if we consult the Experiencer, the pig, we get a different answer. The Experiencer is having fun! It just so happens that the very act of consulting renders the Experiencer mute.

There is something unfair about this. You are trying to make a decision but only listening to an advocate for one side of the argument. And it's not as if the Observer is always right. Imagine a young woman who has a deep romantic and sexual relationship with another woman and gets pleasure and satisfaction from it. But she had a fundamentalist upbringing, and when she talks to her parents, she becomes ashamed of her life.

Or imagine a man who enjoys spending time with his children, his partner, and his friends, but every once in a while he thinks about his career and how others are more successful than him and make more money, and he is upset with himself for his lack of ambition and pledges to spend more time at the office. In each of these cases, the Experiencer is happy and the Observer is not. Is it so obvious that the Observer is correct?

The only fair way to figure out the good life, Gilbert argues, is "duration weighting": see how much time we spend happy and how much time we spend sad, and just add it all up. If you are miserable when you reflect on your life, but you do this reflection for only, say, a couple of hours every week, then this misery shouldn't have much weight. To help us see his side of things, Gilbert writes,

> I can probably get you to at least glimpse the appeal of duration-weighting by asking not which life you would choose for yourself, dear Socrates, but for your child. Would you rather that your child has a life in which she was almost always happy except when she reflected on her life, or the other way around? . . . It's hard to imagine condemning our children to 23 hours of unhappiness every day just so they'll be glad for 1.

I wouldn't have spent so much time on this argument if I didn't think it was worth taking seriously. But I don't fully buy it.

First, getting me to think about my child doesn't do the work Gilbert thinks it does. I'm enough of a motivational pluralist that I wouldn't want to give my child just one hour of pleasure per day. On the other hand, I'd be quite disappointed if either of my sons was a happy couch potato wasting his life. The other name for the Experiencer, following Mill, is the pig. And who wants to have a pig as a child?

Second, I don't think that Socrates and the pig *should* get equal say. As Mill put it, the Experiencer, the pig, knows only its own side of the question. Socrates can reason about the merits of hedonism, which is exactly what we're doing now. And he cares about other people. Being smart can make one vulnerable to destructive ideologies, but still, a pig can only be a pig. If forced to choose, I'd listen to Socrates. He might know something that the pig doesn't. He's the smart one.

After all, you *should* want to have a life that involves projects and plans, one where you connect with others and try to improve their lives. By the metric of duration weighting, a perfectly stoned addict plugged into a never-ending stream of morphine—or someone in Nozick's experience machine, the thought experiment described in the first chapter—lives the best life of all. As does a cheerful sadist, getting immense pleasure from all of the suffering in the world. I don't see it as a controversial moral position to insist that these are not good lives.

There is also a more prosaic reason not to spend the rest of your life in Gilbert's pool. You will probably get tired of it. This is one reason, I would suggest, that having a life of meaning and having a life of pleasure often go together. Long-term difficult projects, for instance, provide opportunities for novelty and excitement; they avoid one of the big problems faced by hedonists: boredom. While the goals of the Observer and the Experiencer can and do often clash, a good life has the promise of satisfying them both.

—

MUCH OF THIS book has been an extended argument that chosen suffering can generate and enhance pleasure, and that it is an essential part of meaningful activities and a meaningful life. And it's often the right thing to do. I'll repeat the quote from Zadie Smith: "It hurts just as much as it is worth." Sometimes pain is a proper acknowledgment of value.

And so, suffering is often a good thing. But not always. Sometimes we overvalue it; sometimes we indulge too much.

To revisit one instance of this, in a previous chapter we discussed how people think there is a logic to suffering, that everything happens for a reason. Even young children are prone to think this way, and it gets enhanced and developed throughout one's life, particularly through exposure to religion. It is not entirely a bad thing to believe. It can be a relief to take unchosen suffering and give it meaning and purpose. It can reassure and soothe us.

But there is a downside here. It can motivate blame. "Everything happens for a reason" implies that people get what they deserve—what goes around comes around. It can lead to a reflexive condemnation of those (including, sometimes, ourselves) who have had bad luck, have become sick, or have been victimized by others. It can also lead to apathy and indifference. If there are no accidents, and everything is ultimately in the service of some higher good, why work so hard to make things better? If discrimination and oppression reflect the workings of a deep plan—the meek shall inherit the earth, after all—why worry about it?

Also, the notion that everything happens for a reason is mistaken, and we shouldn't believe in things that are mistaken. Now, maybe you don't go as far as Richard Dawkins, who has written that the universe exhibits "precisely the properties we should expect if there is, at bottom, no design, no purpose, no evil and no good, nothing but blind, pitiless indifference." But even those who are devout should agree that, at least here on earth, things just don't naturally work out so that people get what they deserve. If there is such a thing as divine justice or karmic retribution, the world we live in is not the place to find it. Instead, the events of human life unfold in a fair and just manner only when individuals and society work to make this happen. We should resist our natural urge to think otherwise.

A further concern was also raised in the previous chapter, having to do with how many of us connect suffering with goodness. We end up judging the merits of an act not just in terms of its intent and consequences but also by considering how much suffering the do-gooder went through. This leads us to discount good acts that don't involve suffering and overrate those that do. And this is pretty foolish. Sometimes, altruistic acts that make the world better also make the altruist happier, and even richer. When people get upset at someone who makes money while improving the lives of others—more upset at him than at someone who does nothing at all—they are discouraging actions that will make the world better.

Finally, chosen suffering can become an end in itself, and can distract from other goods. In *The Body in Pain*, Elaine Scarry talks about artists as "the most authentic class of sufferers," but this isn't a compliment; she worries that their experience and their art "may inadvertently appropriate concern away from others in radical need of assistance." It's more fun to vicariously experience the suffering of fictional people such as Anna Karenina or distant people such as Princess Diana than to engage with the messy individuals around us, who are less interesting, require our attention and effort and resources, and often don't appreciate what we do for them.

In a bracing discussion called "The Banality of Empathy," Namwali Serpell makes a similar argument, quoting Jean-Jacques Rousseau on this point:

> In giving our tears to these fictions, we have satisfied all the rights of humanity without having to give anything more of ourselves; whereas unfortunate people in person would require attention from us, relief, consolation, and work, which would involve us in their pains and would require at least the sacrifice of our indolence.

When made public, certain displays of empathic suffering bring attention, kindness, and love—and, in some circles, a certain authority. We see an extreme version of this on social media, where people often desperately call attention to their own suffering that has been caused by the pain of others. Vicarious suffering might also satisfy a need for human connection. James Dawes writes, "There is a deep satisfaction, a sorrowful joy, that comes from the experience of solidarity in suffering, from sharing one's grief and feeling the weight of another's. Sometimes I think it is a basic human need: for connectedness, for something beyond the existentially impoverished quality of most human interactions."

I'm not as worried as others that vicarious suffering keeps us from helping people in the real world. Still, putting aside any practical concerns, I do share the intuition that there's something repellent about certain acts of vicarious suffering.

The Holocaust scholar Eva Hoffman talks about how in the 1960s there was a fascination with those who had survived the concentration camps, a sort of "depth larceny" whereby affluent Americans would boast about their contact with survivors. She tells of a conversation at a party where she heard one person talking about having a friend who survived Buchenwald and the other crowing in response that one of his neighbors was in Auschwitz. This is narcissistic and disrespectful.

OUR CHOICE TO suffer is not an unmitigated good, then. It has its risks, practical and moral.

But still, chosen suffering—in the right way at the right time in the right doses—adds value to life. I began the book with a defense of motivational pluralism. There are many things we want out of life, and suffering can enhance many of these. Chosen suffering can lead to great pleasure; and it is an essential part of experiences that we deem to be meaningful. It can connect us to others and can be a source of community and love. It reflects deep sentiments of the mind and feelings of the heart.

Our investigation into suffering has also been an exploration of the human condition. The appetite for suffering tells us something important about who we are. At the very least, it shows us that simple theories of what we want are mistaken. We are complicated beings, with a variety of motivations and desires that can be satisfied in surprising ways.

Aldous Huxley makes this point well. His 1932 novel *Brave New World* described a society of stability, control, and drug-induced happiness—a society that sacrificed everything else for the goal of maximizing happiness and pleasure. Near the end of the book, there is a conversation between Mustapha Mond, the representative of the establishment, and John, who has rebelled against the system. Mond argues heatedly for the value of pleasure. He goes on about the neurological interventions being developed to maximize human pleasure, how convenient and easy it all is, and he concludes by saying, "We prefer to do things comfortably."

And John responds, "But I don't want comfort. I want God, I want poetry, I want real danger, I want freedom, I want goodness. I want sin."

There is no better summary of human nature.

ACKNOWLEDGMENTS

I am grateful to so many people for their help. The many thoughtful comments and suggestions that I received when discussing early versions of these ideas in talks, in seminars, and on podcasts have had a profound impact on how I have come to think about pleasure and suffering. As I wrote the book, many scholars—including people I've never met—were kind enough to provide helpful comments and advice when I reached out to them over email. And much of what's here arose from casual conversations with friends—someone would tell a story, or talk about a new finding they'd just read about, or cleverly riff on something I was saying, and I would scramble to write it down so I could use it later.

I know I'm going to forget some names here, and I'm sorry for that, but big thanks to: Ned Block, Max Bloom, Zachary Bloom, Leona Brandwene, Nicholas Christakis, Chaz Firestone, Brett Ford, Deborah Fried, Daniel Gilbert, Sam Harris, Yoel

Inbar, Michael Inzlicht, Julian Jara-Ettinger, Paul Jose, David Kelley, Joshua Knobe, Louisa Lombard, Geoffrey MacDonald, Gregory Murphy, Michael Norton, Gabriele Oettingen, Annie Murphy Paul, Laurie Paul, David Pizarro, Azim Shariff, Tamler Sommers, Amy Starmans, Yaacov Trope, Graeme Wood, Karen Wynn, Dimitris Xygalatas, and Grace Zimmer.

In the early pre-pandemic months of 2020, I finished a complete draft of this book, and I met three times with students and colleagues at Yale to discuss it. I am grateful to the participants in these discussions for their insights: Pinar Aldan, Sophie Arnold, Mario Attie, Jack Beadle, Nicole Betz, Karli Cecil, Vlad Chituc, Joanna Demaree-Cotton, Yarrow Dunham, Brian Earp, Emily Gerdin, Julia Marshall, Laurie Paul, Madeline Reinecke, Alexa Sacchi, Anna-Katrine Sussex, Matti Wilks, Kate Yang, and Katherine Ziska.

A special shout-out to Zachary Bloom, Yarrow Dunham, Frank Keil, Christina Starmans, and Matti Wilks, each of whom gave very helpful written comments on an entire draft.

Daniel Gilbert deserves special mention for nearly persuading me that the whole idea of this book is deeply confused. (He did this once I had finished writing it—thanks, Dan!) And Graeme Wood wins the prize for the weirdest bit of advice; he suggested that the front cover should be a photo of me trussed up with a ball gag.

This is my fifth book with my agent Katinka Matson, and, as always, I'm grateful for her good advice and sharp wit. I am lucky to have her on my side. And it's my second with Denise Oswald, a smart and supportive editor whose wise comments on an earlier version of this manuscript have been invaluable.

Thanks as well to Will Palmer for his excellent copyediting; I doubt that anyone else will ever read this book so closely.

I wrote this book during a transitional period of my life, and I'm grateful to those special people who kept me grounded and had my back, especially Frank Keil, Gregory Murphy, Laurie Paul, Graeme Wood, and my extraordinary sons, Max and Zachary. And lots of love to my family, in Canada and the United States, separated now by the global pandemic, for their kindness and support.

My biggest debt is to my partner, Christina Starmans. Every idea in this book has been shaped by conversations with her and by her penetrating (and often very funny) comments on earlier drafts—if you find sections that are weak, examples that don't work, or jokes that fall flat, those are the parts where I didn't take her advice. Most of this was written in Toronto, where we worked in side-by-side studies, raising our voices to ask questions or report on what we just saw on Twitter, and darting back and forth to try out new ideas or get feedback on what we'd just written. Having Christina at my side made writing this book a delight.

Much of what you have read is about the proper balance of pleasure and purpose, joy and meaning. With Christina I have found the sweet spot. I dedicate this book to her.

NOTES

Preface: The Good Life

xii unpleasant dreams: Antti Revonsuo, "The Reinterpretation of Dreams: An Evolutionary Hypothesis of the Function of Dreaming," *Behavioral and Brain Sciences* 23 (2000): 877–901.

xii even when we daydream: Matthew A. Killingsworth and Daniel T. Gilbert, "A Wandering Mind Is an Unhappy Mind," *Science* 330 (2010): 932.

xiii *The Chronicle of Higher Education* profiled: Tom Bartlett, "Two Famous Academics, 3,000 Fans, $1,500 Tickets," *Chronicle of Higher Education*, April 4, 2019, https://www.chronicle.com/interactives/20190404-peterson.

xiv motivational pluralism: Tyler Cowen, *Stubborn Attachments: A Vision for a Society of Free, Prosperous, and Responsible Individuals* (Stripe Press, 2018), 17.

xv "Genuinely happy individuals are": Mihaly Csikszentmihalyi, *Flow: The Psychology of Optimal Experience* (Harper & Row, 1990), 11.

xvi radical jumps since the 1960s: Emily Esfahani Smith, *The Power of Meaning: Crafting a Life That Matters* (Random House, 2017), 22.

xvi "depressed and severely anxious": Johann Hari, *Lost Connections:*

Uncovering the Real Causes of Depression—and the Unexpected Solutions (Bloomsbury USA, 2018), 11.

xvi "a conspiracy against joy": David Brooks, *The Second Mountain: The Quest for a Moral Life* (Random House, 2019), xxii.

xvi the world has been getting better: Steven Pinker, *Enlightenment Now: The Case for Reason, Science, Humanism, and Progress* (Penguin, 2018).

xviii positive trend in happiness: Steven Pinker, "Enlightenment Wars: Some Reflections on 'Enlightenment Now,' One Year Later," *Quillette*, January 14, 2019, https://quillette.com/2019/01/14/enlightenment-wars-some-reflections-on-enlightenment-now-one-year-later/.

xviii 86 percent of people: Pinker, *Enlightenment Now.*

xviii people underestimate how happy other people are: Pinker, *Enlightenment Now.*

xviii Some countries are happier than others: Ed Diener et al., "Findings All Psychologists Should Know from the New Science on Subjective Well-Being," *Canadian Psychology* 58 (2017): 87–104. To look at the most recent data on happiness across countries, see https://worldhappiness.report.

xix liberals and conservatives: Diener et al., "Findings All Psychologists Should Know."

xix immigrants and native-born citizens: John Helliwell, Richard Layard, and Jeffrey Sachs, *World Happiness Report 2018* (New York: Sustainable Development Solutions Network, 2018), https://worldhappiness.report/ed/2018.

xx some of the claims about this: Helliwell, Layard, and Sachs, *World Happiness Report 2018.*

xx decline in the rate of suicide: "Why Suicide Is Falling Around the World, and How to Bring It Down More," *The Economist*, November 24, 2018, https://www.economist.com/leaders/2018/11/24/why-suicide-is-falling-around-the-world-and-how-to-bring-it-down-more.

xx American suicides have shot upward: Pinker, *Enlightenment Now.*

xx "slow-motion suicide": Brooks, *The Second Mountain.*

xx "Facebook friends in place of neighbors": Hari, *Lost Connections*, 88.

xxi In his book *Tribe*: Sebastian Junger, *Tribe: On Homecoming and Belonging* (Twelve, 2016), 2 and 3.

xxii As Peter Thiel puts it: Peter A. Thiel and Blake Masters, *Zero to One: Notes on Startups, or How to Build the Future* (Broadway Business, 2014), 95 and 96.

xxii "Before I started school striking": Greta Thunberg (@GretaThunberg), Twitter, August 31, 2019, 5:47 p.m., https://twitter.com/GretaThunberg/status/1167916944520908800.

xxiii Viktor Frankl came to a similar conclusion: Joseph B. Fabry, *The Pursuit of Meaning: Viktor Frankl, Logotherapy, and Life* (Harper & Row, 1980).

xxiv the classic Holocaust narrative: Viktor E. Frankl, *Man's Search for Meaning* (Pocket Books, 1973).

xxvi influenced by excellent books: Smith, *The Power of Meaning*; Brock Bastian, *The Other Side of Happiness: Embracing a More Fearless Approach to Living* (Allen Lane, 2018).

1: Suffer

1 "sweet pain": By a blogger named Tom, "Daydreaming @ Mile 110," *Chasing Long*, January 30, 2017, https://chasinglong.blog/2017/01/30/daydreaming-mile-110.

5 "shortage of scarcity": George Ainslie, "Beyond Microeconomics: Conflict Among Interests in a Multiple Self as a Determinant of Value," in *The Multiple Self*, ed. Jon Elster (Cambridge University Press, 1986), 156.

5 an old *Twilight Zone* episode: For the exact dialogue see Wikipedia, s.v. "A Nice Place to Visit," https://en.wikipedia.org/wiki/A_Nice_Place_to_Visit. I also give this example in Paul Bloom, *How Pleasure Works: The New Science of Why We Like What We Like* (Random House, 2010). Similar themes are explored in the final season of the TV show *The Good Place*.

5 a madman who feels pain: David Lewis, "Mad Pain and Martian Pain," in *Philosophical Papers*, vol. 1 (Oxford University Press, 1983).

6 technical definition: Elsa Wuhrman, "Acute Pain: Assessment and Treatment," Medscape, January 3, 2011, https://www.medscape.com/viewarticle/735034.

7 "two sovereign masters": Jeremy Bentham, *An Introduction to the Principles of Morals and Legislation* (Wentworth Press, 2019), 7.

7 "put out of action by a drug": Sigmund Freud, "The Economic Problem of Masochism," in *The Standard Edition of the Complete*

Psychological Works of Sigmund Freud, Volume XIX (1923–1925): The Ego and the Id and Other Works (Hogarth, 1971), 160.

8 "this is nothing": Nikola Grahek, *Feeling Pain and Being in Pain* (MIT Press, 2011), 45.

8 two kinds of analgesics: Grahek, *Feeling Pain and Being in Pain*; Daniel C. Dennett, "Why You Can't Make a Computer That Feels Pain," *Synthese* 38 (1978): 415–56.

9 what pain asymbolia feels like: Grahek, *Feeling Pain and Being in Pain*, 34.

9 "like exes swapping nods": Andrea Long Chu, "The Pink," *n+1* 34 (Spring 2019), https://nplusonemag.com/issue-34/politics/the-pink/.

10 "more 'whoa!' than usual": Robert Wright, *Why Buddhism Is True: The Science and Philosophy of Meditation and Enlightenment* (Simon & Schuster, 2017), 70.

11 fans of horror movies experienced just as much fear: Eduardo B. Andrade and Joel B. Cohen, "On the Consumption of Negative Feelings," *Journal of Consumer Research* 34 (2007): 283–300.

11 anger can be useful: Brett Q. Ford and Maya Tamir, "When Getting Angry Is Smart: Emotional Preferences and Emotional Intelligence," *Emotion* 12 (2012): 685–89.

12 people were shown sad movies: Julian Hanich et al., "Why We Like to Watch Sad Films: The Pleasure of Being Moved in Aesthetic Experiences," *Psychology of Aesthetics, Creativity, and the Arts* 8 (2014): 130–43.

12 while listening to such classical compositions: Ai Kawakami et al., "Relations Between Musical Structures and Perceived and Felt Emotions," *Music Perception: An Interdisciplinary Journal* 30 (2013) 407–17; Liila Taruffi and Stefan Koelsch, "The Paradox of Music-Evoked Sadness: An Online Survey," *PLoS One* 9 (2014): e110490.

12 someone whose heart has recently been broken: Emily Cornett, "Why Do We Enjoy Sad Music? A Review" (unpublished paper, Yale University undergraduate seminar, 2018), cited with permission.

13 Some people with depression: Paul Gilbert et al., "Fears of Compassion and Happiness in Relation to Alexithymia, Mindfulness, and Self-Criticism," *Psychology and Psychotherapy: Theory, Research and Practice* 85 (2012): 374–90.

13 cross-cultural differences: Yuri Miyamoto and Xiaoming Ma, "Dampening or Savoring Positive Emotions: A Dialectical Cultural Script Guides Emotion Regulation," *Emotion* 11 (2011): 1346–57.

13 "Happiness rests in misery": *Laotzu's Taoteching*, trans. Red Pine (Copper Canyon Press, 2009), 116.

13 the mixed nature of the emotions: An Sieun et al., "Two Sides of Emotion: Exploring Positivity and Negativity in Six Basic Emotions Across Cultures," *Frontiers in Psychology* 8 (2017): 610.

14 the first time you get hit in the face: Josh Rosenblatt, *Why We Fight: One Man's Search for Meaning Inside the Ring* (Ecco, 2019), 2.

15 Gilbert's own view: Daniel Gilbert, *Stumbling on Happiness* (Knopf, 2006), 33.

16 "we have a bad habit": Ursula K. Le Guin, "The Ones Who Walk Away from Omelas," *New Dimensions* 3 (1973), https://libcom.org/files/ursula-k-le-guin-the-ones-who-walk-away-from-omelas.pdf.

16 Many researchers in positive psychology: Ed Diener et al., "Findings All Psychologists Should Know from the New Science on Subjective Well-Being," *Canadian Psychology* 58 (2017): 87–104.

16 The words "happiness" and "happy": Anna Wierzbicka, "'Happiness' in Cross-Linguistic & Cross-Cultural Perspective," *Daedalus* 133 (2004): 34–43.

17 a Nazi commandant: Philippa Foot, *Natural Goodness* (Clarendon Press, 2003).

17 Some experimental philosophers: Jonathan Phillips, Luke Misenheimer, and Joshua Knobe, "The Ordinary Concept of Happiness (and Others Like It)," *Emotion Review* 3 (2011): 320–22.

18 Kahneman and his colleagues: Daniel Kahneman and Jason Riis, "Living, and Thinking About It: Two Perspectives on Life," in *The Science of Well-Being*, eds. Felicia Huppert, Nick Baylis, and Barry Keverne (Oxford University Press, 2005): 285–304; Daniel Kahneman and Angus Deaton, "High Income Improves Evaluation of Life but Not Emotional Well-Being," *Proceedings of the National Academy of Sciences* 107 (2010): 16489–93.

18 about three seconds: Marc Wittmann, *Felt Time: The Psychology of How We Perceive Time* (MIT Press, 2016).

18 very boring moments: Dan Gilbert, "Three Pictures of Water: Some Reflections on a Lecture by Daniel Kahneman" (unpublished manuscript, Harvard University, 2008), cited with permission.

19 a survey of a thousand U.S. residents: Kahneman and Deaton, "High Income Improves Evaluation of Life."

20 a poll from 2019: Harvard T.H. Chan School of Public Health, *Life Experiences and Income Inequality in the United States* (NPR/Robert Wood Johnson Foundation/Harvard School of Public Health, 2020), https://www.rwjf.org/en/library/research/2019/12/life-experiences-and-income-inequality-in-the-united-states.html; Christopher Ingraham, "The 1% Are Much More Satisfied with Their Lives than Everyone Else, Survey Finds," *Washington Post*, January 9, 2020, https://www.washingtonpost.com/business/2020/01/09/1-are-much-more-satisfied-with-their-lives-than-everyone-else-survey-finds/.

21 people with more than $10 million: Grant E. Donnelly et al., "The Amount and Source of Millionaires' Wealth (Moderately) Predict Their Happiness," *Personality and Social Psychology Bulletin* 44 (2018): 684–99.

22 "I don't think that people maximize happiness in that sense": Tyler Cowen, "Daniel Kahneman on Cutting Through the Noise," *Conversations with Tyler* podcast, episode 56, December 19, 2018, https://medium.com/conversations-with-tyler/tyler-cowen-daniel-kahneman-economics-bias-noise-167275de691f.

22 "I don't want to be perpetually giddy": Dylan Matthews, "Angus Deaton's Badly Misunderstood Paper on Whether Happiness Peaks at $75,000, Explained," *Vox*, October 12, 2015, https://www.vox.com/2015/6/20/8815813/orange-is-the-new-black-piper-chapman-happiness-study.

23 "Let your belly be full": Andrew George trans., *The Epic of Gilgamesh* (Penguin, 2003).

24 morality is bred in the bone: Paul Bloom, *Just Babies: The Origins of Good and Evil* (Crown, 2013).

24 "watch a hypocrite bleed": Michael Ghiselin, *The Economy of Nature and the Evolution of Sex* (University of California Press, 1974), 247.

25 story of Abraham Lincoln: Cited by Daniel Batson et al., "Where Is the Altruism in the Altruistic Personality?" *Journal of Personality and Social Psychology* 50 (1986): 212–20.

26 psychological hedonism implausible: For reviews, see C. Daniel Batson, *Altruism in Humans* (Oxford University Press, 2011), and Andrew Moore, "Hedonism," *The Stanford Encyclopedia of Philosophy* (Winter 2019 Edition), ed. Edward N. Zalta, https://plato.stanford.edu/archives/win2019/entries/hedonism.

28 Studies of voting patterns: Robert Kurzban, *Why Everyone (Else) Is a Hypocrite: Evolution and the Modular Mind* (Princeton University Press, 2012).

31 experience machine: Robert Nozick, *Anarchy, State, and Utopia* (Basic Books, 1974).

31 "Bye nerd": Philosophy Tube (@PhilosophyTube), Twitter, January 10, 2020.

32 corrupted by a status quo bias: Felipe De Brigard, "If You Like It, Does It Matter If It's Real?" *Philosophical Psychology* 23 (2010): 43–57.

32 a series of surveys on hundreds of individuals: Roy F. Baumeister et al., "Some Key Differences Between a Happy Life and a Meaningful Life" *Journal of Positive Psychology* 8 (2013): 505–16.

34 "The results revealed that happiness": Kathleen D. Vohs, Jennifer L. Aaker, and Rhia Catapano, "It's Not Going to Be That Fun: Negative Experiences Can Add Meaning to Life," *Current Opinion in Psychology* 26 (2019): 11–14.

35 Gallup polled more than 140,000 respondents: Shigehiro Oishi and Ed Diener, "Residents of Poor Nations Have a Greater Sense of Meaning in Life than Residents of Wealthy Nations," *Psychological Science* 25 (2014): 422–30.

35 Since religion also correlates with poverty: For instance, Steve Crabtree, "Religiosity Highest in World's Poorest Nations," Gallup, August 31, 2010, https://news.gallup.com/poll/142727/religiosity-highest-world-poorest-nations.aspx.

36 "more meaningful over time": Adam Alter, "Do the Poor Have More Meaningful Lives?" *New Yorker*, January 24, 2014, https://www.newyorker.com/business/currency/do-the-poor-have-more-meaningful-lives.

37 dream about whatever you want: Alan Watts, "The Dream of Life," Genius, https://genius.com/Alan-watts-the-dream-of-life-annotated.

2: Benign Masochism

40 commercial made for the Olympics: "P&G Thank You, Mom | Pick Them Back Up | Sochi 2014 Olympic Winter Games," January 7, 2014, YouTube video, 2:00, https://www.youtube.com/watch?v=6Ult4t-1NoQ.

40 One of my favorite books: James Elkins, *Pictures and Tears: A History of People Who Have Cried in Front of Paintings* (Routledge, 2005).

40 asked people to watch a sad movie scene: Barbara L. Fredrickson and Robert W. Levenson, "Positive Emotions Speed Recovery from the Cardiovascular Sequelae of Negative Emotions," *Cognition and Emotion* 12 (1998): 191–220.

40 the face associated with orgasm: Susan M. Hughes and Shevon E. Nicholson, "Sex Differences in the Assessment of Pain Versus Sexual Pleasure Facial Expressions," *Journal of Social, Evolutionary, and Cultural Psychology* 2 (2008): 289–98.

41 The authors of a paper in *Science*: Hillel Aviezer, Yaacov Trope, and Alexander Todorov, "Body Cues, Not Facial Expressions, Discriminate Between Intense Positive and Negative Emotions," *Science* 338 (2012): 1225–29.

41 a survey done by Oriana Aragón: Oriana R. Aragón et al., "Dimorphous Expressions of Positive Emotion: Displays of Both Care and Aggression in Response to Cute Stimuli," *Psychological Science* 26 (2015): 259–73.

42 "hedonic flexibility principle": Maxime Taquet et al., "Hedonism and the Choice of Everyday Activities," *Proceedings of the National Academy of Sciences* 113 (2016): 9769–73.

43 "humiliated and abused": Roy F. Baumeister, "Masochism as Escape from Self," *Journal of Sex Research* 25 (1988): 28–59.

43 moral masochism: Sigmund Freud, "The Economic Problem of Masochism," in *The Standard Edition of the Complete Psychological Works of Sigmund Freud, Volume XIX (1923–1925): The Ego and the Id and Other Works,* trans. James Strachey (1964; Hogarth, 1971).

43 "benign masochism": Paul Rozin et al., "Glad to Be Sad, and Other Examples of Benign Masochism," *Judgment and Decision Making* 8 (2013): 439–47.

43 World Sauna Competition: "Sauna Contest Leaves Russian Dead and Champion Finn in Hospital," *Guardian*, August 8, 2010, https://www.theguardian.com/world/2010/aug/08/sauna-championship-russian-dead.

44 "opponent-process" theory: See, for example, Richard L. Solomon, "The Opponent-Process Theory of Acquired Motivation: The Costs of Pleasure and the Benefits of Pain," *American Psychologist* 35 (1980): 691–712.

45 machines that track people's eye movements: R. W. Ditchburn

and B. L. Ginsborg, "Vision with a Stabilized Retinal Image," *Nature* 170 (1952): 36–37.

45 "Because the brain grades on a curve": Indira M. Raman, "Unhappiness Is a Palate-Cleanser," *Nautilus,* March 15, 2018, http://nautil.us/issue/58/self/unhappiness-is-a-palate_cleanser.

46 relative to their expectations: Robb B. Rutledge et al., "A Computational and Neural Model of Momentary Subjective Well-Being," *Proceedings of the National Academy of Sciences* 111 (2014): 12252–57.

46 "pleasant pain": Siri Leknes et al., "The Importance of Context: When Relative Relief Renders Pain Pleasant," *PAIN* 154 (2013): 402–10.

47 like the taste of chocolate: Brock Bastian et al., "The Positive Consequences of Pain: A Biopsychosocial Approach," *Personality and Social Psychology Review* 18 (2014): 256–79.

50 a quirk of how we interpret our own experiences: Daniel Kahneman et al., "When More Pain Is Preferred to Less: Adding a Better End," *Psychological Science* 4 (1993): 401–5.

51 imagine going to the dentist: I gave this example in Paul Bloom, "First-Person Plural," *Atlantic,* November 2008, https://www.theatlantic.com/magazine/archive/2008/11/first-person-plural/307055.

52 "The negative can be an investment": George Ainslie, "Positivity Versus Negativity Is a Matter of Timing," *Behavioral and Brain Sciences* 40 (2017): 16–17.

52 John Wick: Plot summary posted by Nick Riganas on IMDb, https://www.imdb.com/title/tt2911666/plotsummary.

53 the ugliness of some art: Winfried Menninghaus et al., "The Distancing-Embracing Model of the Enjoyment of Negative Emotions in Art Reception," *Behavioral and Brain Sciences* 40 (2017): 1–58.

54 "A whip is a great way": Pat Califia, "Gay Men, Lesbians, and Sex: Doing It Together," *Advocate* 7 (1983): 24–27.

54 in sexual masochism: Roy F. Baumeister, "Masochism as Escape from Self," *Journal of Sex Research* 25 (1988): 28–59.

56 her discussion of torture: Elaine Scarry, *The Body in Pain: The Making and Unmaking of the World* (Oxford University Press, 1985).

56 Control and consent are morally essential: Paul Bloom, "It's Ridiculous to Use Virtual Reality to Empathize with Refugees,"

Atlantic, February 3, 2017, https://www.theatlantic.com/technology/archive/2017/02/virtual-reality-wont-make-you-more-empathetic/515511.

57 "kingdom of hell": Roy F. Baumeister, *Masochism and the Self* (Psychology Press, 2014).

57 involved in BDSM: Juliet Richters et al., "Demographic and Psychosocial Features of Participants in Bondage and Discipline, 'Sadomasochism' or Dominance and Submission (BDSM): Data from a National Survey," *Journal of Sexual Medicine* 5 (2008): 1660–68.

57 if one asks about sexual fantasies: Christian C. Joyal, Amelie Cossette, and Vanessa Lapierre, "What Exactly Is an Unusual Sexual Fantasy?" *Journal of Sexual Medicine* 12 (2015): 328–40.

58 Participants may actually have: Pamela H. Connolly, "Psychological Functioning of Bondage/Domination/Sado-Masochism (BDSM) Practitioners," *Journal of Psychology & Human Sexuality* 18 (2006): 79–120.

58 Consider *Fifty Shades of Grey*: Emma Green, "Consent Isn't Enough: The Troubling Sex of *Fifty Shades*," *Atlantic*, February 10, 2015, https://www.theatlantic.com/entertainment/archive/2015/02/consent-isnt-enough-in-fifty-shades-of-grey/385267/; Gwen Aviles, "'Fifty Shades of Grey' Was the Best-Selling Book of the Decade," NBCNews.com, December 20, 2019, https://www.nbcnews.com/pop-culture/books/fifty-shades-grey-was-best-selling-book-decade-n1105731.

58 "But why, for women especially": Katie Roiphe, "Working Women's Fantasies," *Newsweek*, March 16, 2012, https://www.newsweek.com/working-womens-fantasies-63915.

60 NSSI: For an excellent review, see Matthew K. Nock, "Self-Injury," *Annual Review of Clinical Psychology* 6 (2010): 339–63.

60 finger amputation: Armando R. Favazza, *Bodies Under Siege: Self-Mutilation, Nonsuicidal Self-Injury, and Body Modification in Culture and Psychiatry* (Johns Hopkins University Press, 2011).

60 exorcism: Mark 5:5 (New International Version).

60 In one study of inpatients: Matthew K. Nock and Mitchell J. Prinstein, "Contextual Features and Behavioral Functions of Self-Mutilation Among Adolescents," *Journal of Abnormal Psychology* 114 (2005): 140–46.

61 "As the blood flows": Jennifer Harris, "Self-Harm: Cutting the Bad Out of Me," *Qualitative Health Research* 10 (2000): 164–73.

62 "Galen's friend was overcome": Keith Hopkins, "Novel Evidence for Roman Slavery," *Past and Present* 138 (1993): 3–27.

62 self-punishment in the lab: Brock Bastian, Jolanda Jetten, and Fabio Fasoli, "Cleansing the Soul by Hurting the Flesh: The Guilt-Reducing Effect of Pain," *Psychological Science* 22 (2011): 334–35.

62 shock machine: Yoel Inbar et al., "Moral Masochism: On the Connection Between Guilt and Self-Punishment," *Emotion* 13 (2013): 14–18.

63 advertise positive aspects of ourselves: See, for instance, Kevin Simler and Robin Hanson, *The Elephant in the Brain: Hidden Motives in Everyday Life* (Oxford University Press, 2017).

64 "toques": Stephen Woodman, "In Mexico, Street Vendors Offer Electric Shocks for a Price," *Culture Trip*, March 22, 2018, https://theculturetrip.com/north-america/mexico/articles/in-mexico-street-vendors-offer-electric-shocks-for-a-price.

64 a cry for help: Marilee Strong, *A Bright Red Scream: Self-Mutilation and the Language of Pain* (Penguin, 1999).

64 self-injury as a costly signal: Edward H. Hagen, Paul J. Watson, and Peter Hammerstein, "Gestures of Despair and Hope: A View on Deliberate Self-harm from Economics and Evolutionary Biology," *Biological Theory* 3 (2008): 123–38.

65 their high price is their very point: Geoffrey Miller, *Spent: Sex, Evolution, and Consumer Behavior* (Penguin, 2010). For a dissenting view, see Paul Bloom, "The Lure of Luxury," *Boston Review*, November 2, 2015, http://bostonreview.net/forum/paul-bloom-lure-luxury.

67 Cinnamon Spiced Pecans: Jeff Michaels, "Three Selections from *The Masochist's Cookbook*," *McSweeney's Internet Tendency*, June 5, 2007, https://www.mcsweeneys.net/articles/three-selections-from-the-masochists-cookbook. I used this joke before, in Paul Bloom, *How Pleasure Works: The New Science of Why We Like What We Like* (Random House, 2010).

68 Cicero described: Atul Gawande, "A Queasy Feeling: Why Can't We Cure Nausea?" *New Yorker*, July 5, 1999.

68 morning sickness: Samuel M. Flaxman and Paul W. Sherman,

"Morning Sickness: A Mechanism for Protecting Mother and Embryo," *Quarterly Review of Biology* 75 (2000): 113–48.

69 being bored lowers both your happiness: Roy F. Baumeister et al., "Some Key Differences Between a Happy Life and a Meaningful Life," *Journal of Positive Psychology* 8 (2013): 505–16.

69 a theory of what bores us: For a review, see Erin C. Westgate and Timothy D. Wilson, "Boring Thoughts and Bored Minds: The MAC Model of Boredom and Cognitive Engagement," *Psychological Review* 125 (2018): 689–713.

70 move the baggage claim area: Alex Stone, "Why Waiting Is Torture," *New York Times*, August 18, 2012, https://www.nytimes.com/2012/08/19/opinion/sunday/why-waiting-in-line-is-torture.html.

70 boredom as being akin to pain: Andreas Elpidorou, "The Bright Side of Boredom," *Frontiers in Psychology* 5 (2014): 1245.

71 the wrong strategy: Westgate and Wilson, "Boring Thoughts and Bored Mind."

72 a clever series of studies: Timothy D. Wilson et al., "Just Think: The Challenges of the Disengaged Mind," *Science* 345 (2014): 75–77.

73 "The cetology sections of *Moby Dick*": Andreas Elpidorou, "Boredom in Art," *Behavioral and Brain Sciences* 40 (2017): 25–26.

73 "Jasper Johns is boring": Andreas Elpidorou, "The Quiet Alarm," *Aeon*, July 10, 2015, https://aeon.co/essays/life-without-boredom-would-be-a-nightmare.

74 "Dunbar loved shooting skeet": Joseph Heller, *Catch-22: A Novel* (1961; Simon & Schuster, 1999), 45, 46.

3: An Unaccountable Pleasure

76 the lives of pygmy chimpanzees: Ian McEwan, "Literature, Science, and Human Nature," in *The Literary Animal: Evolution and the Nature of Narrative*, eds. Jonathan Gottschall and David Sloane Wilson (Northwestern University Press, 2005), 11.

76 Only an elephant has a trunk: Steven Pinker, *The Language Instinct: How the Mind Creates Language* (William Morrow, 1994).

76 novel capacities: Steven Pinker and Paul Bloom, "Natural Language and Natural Selection," *Behavioral and Brain Sciences* 13 (1990): 707–27.

78 "Machiavellian intelligence": Richard Byrne and Andrew Whiten, *Machiavellian Intelligence* (Oxford University Press, 1994).

79 surprisingly gifted at appreciating others' thoughts: Paul Bloom, *Descartes' Baby: How the Science of Child Development Explains What Makes Us Human* (Random House, 2005).

80 "the Darwinian treadmill of death": A. D. Nuttall, *Why Does Tragedy Give Pleasure?* (Oxford University Press, 1996), 77.

80 Some monkeys and chimpanzees: For discussion, see Lindsey A. Drayton and Laurie R. Santos, "A Decade of Theory of Mind Research on Cayo Santiago: Insights into Rhesus Macaque Social Cognition," *American Journal of Primatology* 78 (2016): 106–16.

81 how we spend most of our time: Paul Bloom, *How Pleasure Works: The New Science of Why We Like What We Like* (Random House, 2010).

82 Male rhesus monkeys: Robert O. Deaner, Amit V. Khera, and Michael L. Platt, "Monkeys Pay Per View: Adaptive Valuation of Social Images by Rhesus Macaques," *Current Biology* 15 (2005): 543–48.

82 turkey head on a stick. Alex Boese, *Elephants on Acid: And Other Bizarre Experiments* (Pan Macmillan, 2009).

84 kiss your favorite movie star: George Loewenstein, "Anticipation and the Valuation of Delayed Consumption," *Economic Journal* 97 (1987): 666–84.

86 Reality Lite: Bloom, *How Pleasure Works.*

86 the Colosseum: Garrett G. Fagan, *The Lure of the Arena: Social Psychology and the Crowd at the Roman Games* (Cambridge University Press, 2011).

87 Alypius: Fagan, *The Lure of the Arena.*

87 how much they liked horror movies: Mathias Clasen, Jens Kjeldgaard Christiansen, and John A. Johnson, "Horror, Personality, and Threat Simulation: A Survey on the Psychology of Scary Media," *Evolutionary Behavioral Sciences* 14, no. 3 (2018).

87 *Blasted*: Patrick Healy, "Audiences Gasp at Violence; Actors Must Survive It," *New York Times*, November 5, 2008, https://www.nytimes.com/2008/11/06/theater/06blas.html.

87 *1984*: Travis M. Andrews, "Audiences of Broadway's Graphic Portrayal of '1984' Faint and Vomit," *Washington Post*, June 26, 2017, https://www.washingtonpost.com/news/morning-mix/wp/2017/06/26/audiences-of-broadways-graphic-portrayal-of-1984-faint-and-vomit.

87 *This Is Us*: Christine Mattheis, "Your Weekly Cry-Fest Over 'This Is Us' Has Surprising Health Benefits," *Health*, February 23,

2017, https://www.health.com/mind-body/crying-healthy-this-is-us.

88 choose to think about what makes us sad: Matthew A. Killingsworth and Daniel T. Gilbert, "A Wandering Mind Is an Unhappy Mind," *Science* 330 (2010): 932.

89 "more like hell": Jonathan Gottschall, *The Storytelling Animal: How Stories Make Us Human* (Houghton Mifflin Harcourt, 2012).

89 "Where's the baby, Marni?": This is an edited version from Jonathan Gottschall's *The Storytelling Animal*, 35. The original is in Vivian Gussin Paley, *A Child's Work: The Importance of Fantasy Play* (University of Chicago, 2009).

89 "an unaccountable pleasure": David Hume, "Of Tragedy," in *Hume: Selected Essays*, eds. Stephen Copley and Andrew Edgar (Oxford University Press, 2008).

90 features, not bugs: For an excellent review of these issues, see Ellen Winner, *How Art Works: A Psychological Exploration* (Oxford University Press, 2018).

90 "the more I enjoy it!": Eduardo B. Andrade and Joel B. Cohen, "On the Consumption of Negative Feelings," *Journal of Consumer Research* 34 (2007): 283–300.

91 They tested two types of people: Andrade and Cohen. "On the Consumption of Negative Feelings."

91 showed people thirty-eight movies: Julian Hanich et al., "Why We Like to Watch Sad Films: The Pleasure of Being Moved in Aesthetic Experiences," *Psychology of Aesthetics, Creativity, and the Arts* 8 (2014): 130.

92 survey of horror movie fans: Clasen, Christiansen, and Johnson, "Horror, Personality, and Threat Simulation."

92 "the sight of certain things": Aristotle, *The Poetics of Aristotle*, ed. S. H. Butcher (Palala Press, 2016), 22–23.

93 "The delight of tragedy": Samuel Johnson, *Preface to Shakespeare* (1860; Binker North, 2020), 16.

94 "you evil wretches": Plato, *Republic*, Book IV, Section 440a.

94 "Terror is a passion": Edmund Burke, *A Philosophical Enquiry into the Origin of Our Ideas of the Sublime and Beautiful* (Oxford University Press, 1998), 42.

95 very specific tastes: Jennifer L. Barnes, "Fanfiction as Imaginary Play: What Fan-Written Stories Can Tell Us About the Cognitive Science of Fiction," *Poetics* 48 (2015): 69–82.

95 monster in a box: Paul L. Harris et al., "Monsters, Ghosts and Witches: Testing the Limits of the Fantasy-Reality Distinction in Young Children," *British Journal of Developmental Psychology* 9 (1991): 105–23.

96 soup from a brand-new bedpan: Paul Rozin, Linda Millman, and Carol Nemeroff, "Operation of the Laws of Systematic Magic in Disgust and Other Domains," *Journal of Personality and Social Psychology* 50 (1986): 703–12.

96 the mind works on two tracks: Tamar Szabó Gendler, "Alief in Action (and Reaction)," *Mind & Language* 23 (2008): 552–85.

97 112,000 plots: David Robinson, "Examining the Arc of 100,000 Stories: A Tidy Analysis," *Variance Explained* (blog), n.d., http://varianceexplained.org/r/tidytext-plots/.

98 six main categories: Andrew J. Reagan et al., "The Emotional Arcs of Stories Are Dominated by Six Basic Shapes," *EPJ Data Science* 5 (2016): 31. For discussion, see Adrienne LaFrance, "The Six Main Arcs in Storytelling, as Identified by an A.I," *Atlantic*, July 12, 2016, https://www.theatlantic.com/technology/archive/2016/07/the-six-main-arcs-in-storytelling-identified-by-a-computer/490733/.

99 cognitive science approaches to literary universals: Patrick Colm Hogan, *The Mind and Its Stories: Narrative Universals and Human Emotion* (Cambridge University Press, 2003).

99 *present a formidable obstacle*: Aaron Sorkin, "Intention & Obstacle," MasterClass, https://www.masterclass.com/classes/aaron-sorkin-teaches-screenwriting/chapters/intention-obstacle-11ba8c15-7856-490d-85bb-eb0601e02c55#.

99 even if one knows the outcome: Richard J. Gerrig, "Suspense in the Absence of Uncertainty," *Journal of Memory and Language* 28 (1989): 633–48.

101 an evolved motivation to practice: See, for example, Anthony D. Pellegrini, Danielle Dupuis, and Peter K. Smith, "Play in Evolution and Development," *Developmental Review* 27 (2007): 261–76.

102 "the tough mind's way of coping": Stephen King, *Danse Macabre* (Everest House, 1981), 13, 335. My own thinking on this has changed, and so this is similar but not identical to the account I develop in Bloom, *How Pleasure Works*.

102 "you can never have too much insurance": Jerry Fodor, *In Critical Condition: Polemical Essays on Cognitive Science and the Philosophy of Mind* (MIT Press, 1998), 212.

104 positive fantasies can often be bad for you: Gabriele Oettingen and A. Timur Sevincer, "Fantasy About the Future as Friend and Foe," in *The Psychology of Thinking About the Future*, eds. Gabriele Oettingen et al. (Guilford, 2018).

105 what people search for on pornographic websites: Ogi Ogas, Sai Gaddam, and Andrew J. Garman, *A Billion Wicked Thoughts* (Penguin, 2011), and Seth Stephens-Davidowitz and Andrés Pabon, *Everybody Lies: Big Data, New Data, and What the Internet Can Tell Us About Who We Really Are* (Dey Street, 2017).

106 cartoon pornography: Stephens-Davidowitz and Pabon, *Everybody Lies.*

106 "teenage brothers and sisters": Steven Pinker, *How the Mind Works* (Penguin UK, 2003), 455.

107 one-quarter of female searches: Stephens-Davidowitz and Pabon, *Everybody Lies.*

107 rape fantasies: Joseph W. Critelli and Jenny M. Bivona, "Women's Erotic Rape Fantasies: An Evaluation of Theory and Research," *Journal of Sex Research* 45 (2008): 57–70. For discussion, see Matthew Hudson, "Why Do Women Have Erotic Rape Fantasies?" *Psychology Today*, May 29, 2008, https://www.psychologytoday.com/us/blog/psyched/200805/why-do-women-have-erotic-rape-fantasies.

108 Deep dives into descriptions of these fantasies: Critelli and Bivona, "Women's Erotic Rape Fantasies."

109 "an exaggerated and caricatured morality": David A. Pizarro and Roy F. Baumeister, "Superhero Comics as Moral Pornography," in *Our Superheroes, Ourselves*, ed. Robin Rosenberg (Oxford University Press, 2013), 29.

110 areas of the brain associated with pleasure and reward: Alan G Sanfey et al., "The Neural Basis of Economic Decision-Making in the Ultimatum Game," *Science* 300 (2003): 1755–58.

110 pleasure is grounded in sound evolutionary logic: For discussion, see Paul Bloom, *Just Babies: The Origins of Good and Evil* (Crown, 2013).

111 the negative is more powerful than the positive: Paul Rozin and Edward B. Royzman, "Negativity Bias, Negativity Dominance, and Contagion," *Personality and Social Psychology Review* 5 (2001): 296–320.

112 For many examples of this: Jon Ronson, *So You've Been Publicly Shamed* (Riverhead, 2016).

112 group responses to individual transgressions: Paul Bloom and Matthew Jordan, "Are We All 'Harmless Torturers' Now?" *New York Times*, August 9, 2018, https://www.nytimes.com/2018/08/09/opinion/are-we-all-harmless-torturers-now.html.

4: Struggle

115 How much do you have to pay: Edward L. Thorndike, "Valuations of Certain Pains, Deprivations, and Frustrations," *Pedagogical Seminary and Journal of Genetic Psychology* 51 (1937): 227–39.

117 an important point about our psychologies: Paul Bloom, *Just Babies: The Origins of Good and Evil* (Crown, 2013).

117 self-harm and harm to others: Molly J. Crockett et al., "Harm to Others Outweighs Harm to Self in Moral Decision Making," *Proceedings of the National Academy of Sciences* 111 (2014): 17320–25.

118 talk about our everyday efforts in economic terms: Wouter Kool and Matthew Botvinick, "Mental Labour," *Nature Human Behaviour* 2 (2018): 899–908.

118 *effort*—technically defined as: Michael Inzlicht, Amitai Shenhav, and Christopher Y. Olivola, "The Effort Paradox: Effort Is Both Costly and Valued," *Trends in Cognitive Sciences* 22 (2018): 337–49.

119 In 1890, William James: William James, *The Principles of Psychology* (Macmillan: 1890), 455.

120 Effort taxes the body and the soul: Inzlicht, Shenhav, and Olivola, "The Effort Paradox."

122 this cost factors into how we make sense: Julian Jara-Ettinger et al., "The Naïve Utility Calculus: Computational Principles Underlying Commonsense Psychology," *Trends in Cognitive Sciences* 20 (2016): 589–604.

122 "paradox of choice": Barry Schwartz, *The Paradox of Choice: Why More Is Less* (Ecco, 2004).

122 the extremes of mental effort: Tsuruko Arai, *Mental Fatigue* (PhD diss., Teachers College, Columbia University, 1912). This work is cited in Robert Kurzban et al., "An Opportunity Cost Model of Subjective Effort and Task Performance," *Behavioral and Brain Sciences* 36 (2013): 661–79.

123 researchers replicated Arai's findings: Zelma Langdon Huxtable et al., "A Re-Performance and Re-Interpretation of the Arai Experiment in Mental Fatigue with Three Subjects," *Psychological Monographs* 59, no. 5 (1945), 52.

123 intelligence is related to all sorts of good things: Steven Pinker, *The Better Angels of Our Nature: Why Violence Has Declined* (Penguin, 2012).

124 Deficits in self-control: Walter Mischel, *The Marshmallow Test: Understanding Self-Control and How to Master It* (Random House, 2014).

124 Surveys of productive people. Daniel H. Pink, *When: The Scientific Secrets of Perfect Timing* (Penguin, 2019).

125 "deep work": Cal Newport, *Deep Work: Rules for Focused Success in a Distracted World* (Hachette, 2016).

126 actually is a lot like a muscle: Roy F. Baumeister, Dianne M. Tice, and Kathleen D. Vohs, "The Strength Model of Self-Regulation: Conclusions from the Second Decade of Willpower Research," *Perspectives on Psychological Science* 13 (2018): 141–45.

126 Being high or low in willpower: Brent W. Roberts et al., "What Is Conscientiousness and How Can It Be Assessed?" *Developmental Psychology* 50 (2014): 1315–30.

126 One piece of advice they offer: Roy F. Baumeister and John Tierney, *Willpower: Rediscovering the Greatest Human Strength* (Penguin Books, 2011).

126 I saw that Barack Obama: Michael Lewis, "Obama's Way," *Vanity Fair*, September 11, 2012, https://www.vanityfair.com/news/2012/10/michael-lewis-profile-barack-obama.

127 "Obama Wears Boring Suits": Katherine Mangu-Ward, "Obama Wears Boring Suits So He Won't Tweet Pictures of His Penis," *Reason*, September 14, 2012, https://reason.com/2012/09/14/obama-wears-boring-suits-so-he-wont-twee/.

128 "opportunity cost": Kurzban et al., "An Opportunity Cost Model."

129 "the effort paradox": Inzlicht, Shenhav, and Olivola, "The Effort Paradox."

130 Tom Sawyer: For discussion, see Dan Ariely, George Loewenstein, and Drazen Prelec, "Tom Sawyer and the Construction of Value," *Journal of Economic Behavior & Organization* 60 (2006): 1–10.

130 instant cake mixes: Michael I. Norton, Daniel Mochon, and Dan Ariely, "The IKEA Effect: When Labor Leads to Love," *Journal of Consumer Psychology* 22 (2012): 453–60.

131 "the Ikea effect": Norton, Mochon, and Ariely, "The IKEA Effect."

131 this association between effort and value: Justin Kruger et al., "The Effort Heuristic," *Journal of Experimental Social Psychology* 40 (2004): 91–98.

132 Rats: Inzlicht, Shenhav, and Olivola, "The Effort Paradox."

132 Ants: Tomer J. Czaczkes et al., "Greater Effort Increases Perceived Value in an Invertebrate," *Journal of Comparative Psychology* 132 (2018): 200–209.

134 makes for a good game: For a popular treatment, see Jane McGonigal Read, *Reality Is Broken: Why Games Make Us Better and How They Can Change The World* (Penguin, 2011).

134 *atelic* activities: Kieran Setiya, *Midlife: A Philosophical Guide* (Princeton, 2017).

136 flow: Mihaly Csikszentmihalyi, *Flow: The Psychology of Optimal Experience* (Harper & Row, 1990).

138 surveys on Americans and Germans: Jeanne Nakamura and Mihaly Csikszentmihalyi, "The Concept of Flow," in *Flow and the Foundations of Positive Psychology: The Collected Works of Mihaly Csikszentmihalyi*, by Mihaly Csikszentmihalyi (Springer, 2014).

139 A poll from Gallup: Summarized by Johann Hari in *Lost Connections: Uncovering the Real Causes of Depression—and the Unexpected Solutions* (Bloomsbury USA, 2018).

140 bullshit jobs: David Graeber and Albertine Cerutti, *Bullshit Jobs* (Simon & Schuster, 2018).

140 Some are associated with meaning: "The Most and Least Meaningful Jobs," PayScale, https://www.payscale.com/data-packages/most-and-least-meaningful-jobs. See also Adam Grant, "Three Lies About Meaningful Work," LinkedIn, May 7, 2015, https://www.linkedin.com/pulse/three-lies-meaningful-work-adam-grant.

141 part of the healing process: Amy Wrzesniewski and Jane E. Dutton, "Crafting a Job: Revisioning Employees as Active Crafters of Their Work," *Academy of Management Review* 26 (2001): 179–201.

141 "helping put a man on the moon": Emily Esfahani Smith, *The Power of Meaning: Crafting a Life That Matters* (Random House, 2017).

141 "*a good life*": Nakamura and Csikszentmihalyi, "The Concept of Flow."

142 Eichmann "probably experienced flow": Csikszentmihalyi, *Flow*, 231.

5: Meaning

145 an excellent article: George Loewenstein, "Because It Is There: The Challenge of Mountaineering . . . for Utility Theory," *Kyklos* 52 (1999): 315–43.

145 "little more psychological insight": Loewenstein, "Because It Is There," 315.

145 his actual views were sophisticated: *The Stanford Encyclopedia of Philosophy*, s.v. "Jeremy Bentham" (Summer 2019 Edition), ed. Edward N. Zalta, plato.stanford.edu/archives/sum2019/entries/bentham.

145 "unrelenting misery": Loewenstein, "Because It Is There," 318.

147 curiosity about one's own capacities: Loewenstein, "Because It Is There," 324.

148 reflect your self-doubt: Ronit Bodner and Drazen Prelec, "The Emergence of Private Rules in a Self-Signaling Model," *International Journal of Psychology* 31 (1996): 3652.

151 the transformative power of such an experience: For discussion, see Laurie A. Paul, *Transformative Experience* (Oxford University Press, 2014).

151 "All we hear of ISIS": Joyce Carol Oates (@JoyceCarolOates), Twitter, November 22, 2015, 2:28 p.m.

152 Oates's defense: Ross Douthat, "The Joy of ISIS," *New York Times*, November 23, 2015, https://www.nytimes.com/2015/11/24/books/joyce-carol-oates-celebratory-joyous-islamic-state-twitter.html.

152 a book based on interviews: See Graeme Wood, *The Way of the Strangers: Encounters with the Islamic State* (Random House, 2019).

152 the appeal of National Socialism: George Orwell, "Review of *Mein Kampf*, by Adolph Hitler, unabridged translation," *New English Weekly*, March 21, 1940.

153 *Call of Duty*: Jeff Grubb, "Call of Duty: Modern Warfare Sales Up Big over Black Ops 4," *VentureBeat*, February 20, 2020, https://venturebeat.com/2020/02/06/activision-blizzard-call-of-duty/.

153 a *New Yorker* interview: Michael Shulman, "Adam Driver, the Original Man," *New Yorker*, October 21, 2019, https://www.newyorker.com/magazine/2019/10/28/adam-driver-the-original-man. Thanks to Julian Jara-Ettinger for providing this example.

154 nine hundred employed women: Daniel Kahneman et al., "A Survey Method for Characterizing Daily Life Experience: The Day Reconstruction Method," *Science* 306 (2004): 1776–80.

154 when a child is born: Richard E. Lucas, "Reexamining Adaptation and the Set Point Model of Happiness: Reactions to Changes in Marital Status," *Journal of Personality and Social Psychology* 84 (2003): 527–39. Maike Luhmann, "Subjective Well-Being and Adaptation to Life Events: A Meta-Analysis," *Journal of Personality and Social Psychology* 102 (2012): 592–615.

154 a drop in marital satisfaction: Jean M. Twenge, "Parenthood and Marital Satisfaction: A Meta-Analytic Review," *Journal of Marriage and Family* 65 (2003): 574–83.

154 "The only symptom of empty nest syndrome": Chuck Leddy, "Money, Marriage, Kids," *Harvard Gazette*, February 21, 2013, https://news.harvard.edu/gazette/story/2013/02/money-marriage-kids.

155 children provoke a couple's most frequent arguments: Jennifer Senior, *All Joy and No Fun: The Paradox of Modern Parenthood* (HarperCollins, 2014), 49.

155 happiness hit is worse for some people: Katherine S. Nelson et al., "In Defense of Parenthood: Children Are Associated with More Joy than Misery," *Psychological Science* 24 (2013): 3–10.

155 childcare policies: Jennifer Glass, Robin W. Simon, and Matthew A. Andersson, "Parenthood and Happiness: Effects of Work-Family Reconciliation Policies in 22 OECD Countries," *American Journal of Sociology* 122 (2016): 886–929.

156 Our memory is selective: Senior, *All Joy and No Fun*, 256–57.

156 Kieran Setiya expands on this point: Kieran Setiya, *Midlife: A Philosophical Guide* (Princeton, 2017).

157 "How often, if at all": Nelson et al., "In Defense of Parenthood," 3–10.

157 in the study by Baumeister and his colleagues: Roy F. Baumeister et al., "Some Key Differences Between a Happy Life and a Meaningful Life," *Journal of Positive Psychology* 8 (2013): 505–16.

158 "a strange admixture": Zadie Smith, "Joy," *New York Review of Books*, January 10, 2013, https://www.nybooks.com/articles/2013/01/10/joy.

158 only one serious philosophical problem: Albert Camus, *The Myth of Sisyphus* (Penguin, 2013).

159 "the unexamined life": Casey Woodling, in "What Is the Meaning

of Life?" *Philosophy Now*, 2020, https://philosophynow.org/issues/59/What_Is_The_Meaning_Of_Life.

160 American Freshman Survey: Emily Esfahani Smith, *The Power of Meaning: Crafting a Life That Matters* (Random House, 2017).

161 "'Forty-two!' yelled Loonquawl": Douglas Adams, *The Hitchhiker's Guide to the Galaxy Omnibus: A Trilogy in Five Parts*, vol. 6 (1979; Macmillan, 2017).

162 "The meaning of life is not being dead": Tim Bale, in "What Is the Meaning of Life?" *Philosophy Now*, 2020, https://philosophynow.org/issues/59/What_Is_The_Meaning_Of_Life.

163 nicely expressed by Viktor Frankl: Viktor E. Frankl, *Man's Search for Meaning* (Pocket Books, 1973), 113.

165 Emily Esfahani Smith talks about: Smith, *The Power of Meaning*, 40–41.

166 other, related proposals: Frank Martela and Michael F. Steger, "The Three Meanings of Meaning in Life: Distinguishing Coherence, Purpose, and Significance," *Journal of Positive Psychology* 11 (2016): 531–45; Michael F. Steger, "Meaning in Life," in *The Oxford Handbook of Positive Psychology*, eds. by Shane J. Lopez and C. R. Snyder (Oxford; 2009).

166 "Beyond Bentham": George Loewenstein and Niklas Karlsson, "Beyond Bentham: The Search for Meaning," *JDM Newsletter*, June 2002, http://www.decisionsciencenews.com/sjdm/newsletters/02-jun.pdf.

170 tended to be on the extremes: Sean C. Murphy and Brock Bastian, "Emotionally Extreme Life Experiences Are More Meaningful," *Journal of Positive Psychology* 15, no. 11 (2019): 1–12.

170 "loss of body parts": Loewenstein, "Because It Is There."

171 six-hour layover in the Budapest airport: Anat Keinan and Ran Kivetz, "Productivity Orientation and the Consumption of Collectable Experiences," *Journal of Consumer Research* 37 (2011): 935–50.

173 "I did not only talk of the future": Frankl, *Man's Search for Meaning*, 104.

6: Sacrifice

175 Hindu festival: Dimitris Xygalatas et al., "Extreme Rituals Promote Prosociality," *Psychological Science* 24 (2013): 1602–5.

176 psychological foundations of religion are part of human nature:

Konika Banerjee and Paul Bloom, "Would Tarzan Believe in God? Conditions for the Emergence of Religious Belief," *Trends in Cognitive Sciences* 17 (2013): 7–8; Paul Bloom, "Religion Is Natural," *Developmental Science* 10 (2007): 147–51.

177 "bring out our inner bee": Jonathan Haidt, "Moral Psychology and the Misunderstanding of Religion," *Edge*, September 21, 2007, https://www.edge.org/conversation/jonathan_haidt-moral-psychology-and-the-misunderstanding-of-religion. See also Jonathan Haidt, *The Righteous Mind: Why Good People Are Divided by Politics and Religion* (Vintage, 2012).

177 which societies will last the longest: Ara Norenzayan and Azim F. Shariff, "The Origin and Evolution of Religious Prosociality," *Science* 322 (2008): 58–62.

177 synchrony brings people together: Scott S. Wiltermuth and Chip Heath, "Synchrony and Cooperation," *Psychological Science* 20 (2009): 1–5.

178 high-pain rituals: Xygalatas et al., "Extreme Rituals Promote Prosociality."

179 self-conscious attempts to use rituals: For example, see Alain de Botton, *Religion for Atheists: A Non-Believer's Guide to the Uses of Religion* (Vintage, 2012).

179 whipped by belts: Christopher M. Kavanagh et al., "Positive Experiences of High Arousal Martial Arts Rituals Are Linked to Identity Fusion and Costly Pro-Group Actions," *European Journal of Social Psychology* 49 (2019): 461–81.

180 James Costello: Konika Banerjee and Paul Bloom, "Does Everything Happen for a Reason?" *New York Times*, October 17, 2014, https://www.nytimes.com/2014/10/19/opinion/sunday/does-everything-happen-for-a-reason.html.

181 "psychological immune system": Dan T. Gilbert et al., "Immune Neglect: A Source of Durability Bias in Affective Forecasting," *Journal of Personality and Social Psychology* 75 (1998): 617–38. The Moreese Bickham anecdote is from Dan T. Gilbert, "The Surprising Science of Happiness," TED video, 20:52, February 2004, https://www.ted.com/talks/dan_gilbert_the_surprising_science_of_happiness/transcript.

181 "I wouldn't change anything": Laurie Santos, "The Unhappy Millionaire," *The Happiness Lab* (podcast), https://www.happinesslab.fm/season-1-episodes/the-unhappy-millionaire.

182 give meaning to various life events: Konika Banerjee and Paul Bloom, "Why Did This Happen to Me? Religious Believers' and Non-Believers' Teleological Reasoning About Life Events," *Cognition* 133 (2014): 277–303; Konika Banerjee and Paul Bloom, "'Everything Happens for a Reason': Children's Beliefs About Purpose in Life Events," *Child Development* 86 (2015): 503–18.

183 New Testament: Hebrews 12:7–11 (New International Version).

184 one's relationship with Christ: Brian Pizzalato, "St. Paul Explains the Meaning of Suffering," Catholic News Agency, https://www.catholicnewsagency.com/resources/sacraments/anointing-of-the-sick/st-paul-explains-the-meaning-of-suffering.

184 "the boundless price of our redemption": Pope John Paul II, 1984 address quoted by Ariel Glucklich in *Sacred Pain: Hurting the Body for the Sake of the Soul* (Oxford University Press, 2001), 18.

184 "the rebel soul": C. S. Lewis, *The Problem of Pain* (Harper, 2015), 93–94.

185 "Dr. McCullough said": Ted Chiang, "Omphalos," in *Exhalation: Stories* (Knopf, 2019), 262. (This story is a clever critique of the ideas posited in the 1857 book by Philip Henry Gosse, *Omphalos: An Attempt to Untie the Geological Knot.*)

186 "I wish there were no such thing as anesthesia!": William Henry Atkinson, quoted by Ariel Glucklich in *Sacred Pain*, 184.

186 high desire for pain in S&M sessions: Martin S. Weinberg, Colin J. Williams, and Charles Moser, "The Social Constituents of Sadomasochism," *Social Problems* 31 (1984): 379–389. Recounted in Daniel Bergner, *The Other Side of Desire: Four Journeys into the Far Realms of Lust and Longing* (Penguin, 2009).

187 "Everyone knows that fasting": Lewis, *The Problem of Pain*, 112.

187 "moral weeping": Tom Lutz, *Crying: The Natural and Cultural History of Tears* (Norton, 2001), 11.

188 "It hurts just as much as it is worth": Zadie Smith, "Joy," *New York Review of Books*, January 10, 2013, https://www.nybooks.com/articles/2013/01/10/joy.

188 Daniel Pallotta: The story is told in George E. Newman and Daylian M. Cain's "Tainted Altruism: When Doing Some Good Is Evaluated as Worse than Doing No Good at All," *Psychological Science* 25 (2014): 648–55.

189 "tainted altruism": Newman and Cain, "Tainted Altruism."

190 Suppose you have a friend who is sick: Christopher Y. Olivola

and Eldar Shafir, "The Martyrdom Effect: When Pain and Effort Increase Prosocial Contributions," *Journal of Behavioral Decision Making* 26 (2013): 91–105.

191 commencement speech: John Roberts, commencement speech at Cardigan Mountain School, Canaan, New Hampshire, June 3, 2017; see transcript in "'I Wish You Bad Luck': Read Supreme Court Justice John Roberts' Unconventional Speech to His Son's Graduating Class," *Time*, July 5, 2017, https://time.com/4845150/chief-justice-john-roberts-commencement-speech-transcript/.

191 "anti-fragile": Nassim Nicholas Taleb, *Antifragile: Things That Gain from Disorder* (Random House, 2012).

191 "The key to healthy psychological functioning is exposure": Brock Bastian, *The Other Side of Happiness: Embracing a More Fearless Approach to Living* (Penguin, 2018), 95.

192 put one of their hands in ice-cold water: Mark D. Seeley et al., "An Upside to Adversity? Moderate Cumulative Lifetime Adversity Is Associated with Resilient Responses in the Face of Controlled Stressors," *Psychological Science* 24 (2013): 1181–89.

192 This one focused on kindness: Daniel Lim and David DeSteno, "Suffering and Compassion: The Links Among Adverse Life Experiences, Empathy, Compassion, and Prosocial Behavior," *Emotion* 16 (2016): 175–82.

193 other work suggesting that poor people: Ana Guinote et al., "Social Status Modulates Prosocial Behavior and Egalitarianism in Preschool Children and Adults," *Proceedings of the National Academy of Sciences* 112 (2015): 731–36.

193 how groups of people respond to disasters: Rebecca Solnit, *A Paradise Built in Hell: The Extraordinary Communities That Arise in Disaster* (Penguin, 2010), 8.

193 Brock Bastian and his colleagues: Brock Bastian, Jolanda Jetten, and Laura J. Ferris, "Pain as Social Glue: Shared Pain Increases Cooperation," *Psychological Science* 25 (2014): 2079–85.

194 a Taoist story: This version, titled "Maybe," is from *John Suler's Zen Stories to Tell Your Neighbors*, http://truecenterpublishing.com/zenstory/maybe.html.

196 people can be resilient: George A. Bonanno, "Loss, Trauma, and Human Resilience: Have We Underestimated the Human Capacity to Thrive After Extremely Aversive Events?" *American Psychologist* 59 (2004): 20–28.

196 in a recent review article: Johanna Ray Vollhardt, "Altruism Born of Suffering and Prosocial Behavior Following Adverse Life Events: A Review and Conceptualization," *Social Justice Research* 22, no. 1 (2009): 53–97.

198 a "natural experiment": Anthony D. Mancini, Heather L. Littleton, and Amie E. Grills, "Can People Benefit from Acute Stress? Social Support, Psychological Improvement, and Resilience After the Virginia Tech Campus Shootings," *Clinical Psychological Science* 4 (2016): 401–17.

200 survivorship bias: Jordan Ellenberg, *How Not to Be Wrong: The Power of Mathematical Thinking* (Penguin, 2014).

202 post-traumatic growth: For example, see Eranda Jayawickreme and Laura E. R. Blackie, "Post-Traumatic Growth as Positive Personality Change: Evidence, Controversies and Future Directions," *European Journal of Personality* 4 (2014): 312–31.

202 "People develop new understandings of themselves": Richard Tedeschi, quoted by Lorna Collier in "Growth After Trauma," *Monitor on Psychology* 47, no. 10 (November 2016), 48, https://www.apa.org/monitor/2016/11/growth-trauma.

202 improvement in five areas: For example, see Kanako Taku et al., "The Factor Structure of the Posttraumatic Growth Inventory: A Comparison of Five Models Using Confirmatory Factor Analysis," *Journal of Traumatic Stress* 21 (2008): 158–64.

203 a recent meta-analysis: Judith Mangelsdorf, Michael Eid, and Maike Luhmann, "Does Growth Require Suffering? A Systematic Review and Meta-Analysis on Genuine Posttraumatic and Postecstatic Growth," *Psychological Bulletin* 145 (2019): 302–38.

7: Sweet Poison

205 "the lucky ones": Richard Dawkins, *Unweaving the Rainbow: Science, Delusion and the Appetite for Wonder* (Houghton Mifflin, 1998), 1.

206 "fallen angels": Robert Ardrey, *African Genesis* (Collins, 1961), 245–46.

207 I've argued directly against this theistic analysis: Paul Bloom, "Did God Make These Babies Moral?" *New Republic,* January 13, 2014, https://newrepublic.com/article/116200/moral-design-latest-form-intelligent-design-its-wrong.

208 "We are happier when we are healthy": Steven Pinker, *How the Mind Works* (Penguin UK, 2003), 389.

209 "Modern life is full of emotional reactions": Robert Wright, *Why Buddhism Is True: The Science and Philosophy of Meditation and Enlightenment* (Simon & Schuster, 2017), 36.

212 emotions such as empathy are too biased and innumerate: Paul Bloom, *Against Empathy: The Case for Rational Compassion* (Random House, 2017).

213 It's that there is a correlation: Roy F. Baumeister et al., "Some Key Differences Between a Happy Life and a Meaningful Life," *Journal of Positive Psychology* 8 (2013): 505–16.

213 the relationship between pleasure and meaning: Veronika Huta and Richard M. Ryan, "Pursuing Pleasure or Virtue: The Differential and Overlapping Well-Being Benefits of Hedonic and Eudaimonic Motives," *Journal of Happiness Studies* 11 (2010): 735–62.

214 one can screw up being happy: Brett Q. Ford et al., "Culture Shapes Whether the Pursuit of Happiness Predicts Higher or Lower Well-Being," *Journal of Experimental Psychology: General* 144 (2015): 1053–62.

214 The people who highly agree with such items: Brett Q. Ford et al., "Desperately Seeking Happiness: Valuing Happiness Is Associated with Symptoms and Diagnosis of Depression," *Journal of Social and Clinical Psychology* 33 (2014): 890–905.

215 listening to Stravinsky's *Rite of Spring*: Jonathan W. Schooler, Dan Ariely, and George Loewenstein, "The Pursuit and Monitoring of Happiness Can Be Self-Defeating," *Psychology and Economics* (2003): 41–70.

215 Focusing on happiness does seem to have a bad effect: Iris B. Mauss et al., "Can Seeking Happiness Make People Unhappy? Paradoxical Effects of Valuing Happiness," *Emotion* 11 (2011): 807–15.

215 Maybe when you pursue happiness: Brett Q. Ford and Iris B. Mauss, "The Paradoxical Effects of Pursuing Positive Emotion," in *Positive Emotion: Integrating the Light Sides and Dark Sides*, eds. June Gruber and Judith Tedlie Moskowitz (Oxford University Press, 2014).

215 pursuing extrinsic goals: Tim Kasser and Richard M. Ryan, "Further Examining the American Dream: Differential Correlates of Intrinsic and Extrinsic Goals," *Personality and Social Psychology Bulletin* 22 (1996): 280–87. For the meta-analysis, see Helga Dittmar et al., "The Relationship Between Materialism

and Personal Well-Being: A Meta-Analysis," *Journal of Personality and Social Psychology* 107 (2014): 879–924.

216 cross-cultural research finds: Ford et al., "Culture Shapes Whether the Pursuit of Happiness."

216 "Hey Jude": Marc Wittmann, *Felt Time: The Psychology of How We Perceive Time* (MIT Press, 2016).

217 this view has some sharp defenders: Dan Gilbert, "Three Pictures of Water: Some Reflections on a Lecture by Daniel Kahneman" (unpublished manuscript, Harvard University, 2088). I am grateful to Dan for sending me this paper, and for our many discussions of these matters. And I should add that he no longer fully holds the view that I am currently arguing against.

217 "I was very interested in": Tyler Cowen, "Daniel Kahneman on Cutting Through the Noise," *Conversations with Tyler* podcast, episode 56, December 19, 2018, https://medium.com/conversations-with-tyler/tyler-cowen-daniel-kahneman-economics-bias-noise-167275de691f.

217 "I don't want to be perpetually giddy": Dylan Matthews, "Angus Deaton's Badly Misunderstood Paper on Whether Happiness Peaks at $75,000, Explained," *Vox*, October 12, 2015, https://www.vox.com/2015/6/20/8815813/orange-is-the-new-black-piper-chapman-happiness-study.

217 "It is better to be": John Stuart Mill, *Utilitarianism* (Coventry House Publishing, 2017), 15–16.

222 "pitiless indifference": Richard Dawkins, *River Out of Eden: A Darwinian View of Life* (Basic Books, 2008), 133.

222 "the most authentic class of sufferers": Elaine Scarry, *The Body in Pain: The Making and Unmaking of the World* (Oxford University Press, 1987), 11.

223 "In giving our tears to these fictions": Namwali Serpell, "The Banality of Empathy," *New York Review of Books*, March 2, 2019, https://www.nybooks.com/daily/2019/03/02/the-banality-of-empathy/.

223 "There is a deep satisfaction": James Dawes, *Evil Men* (Harvard University Press, 2013), 208.

224 "depth larceny": Eva Hoffman, *After Such Knowledge: Memory, History, and the Legacy of the Holocaust* (PublicAffairs, 2005).

225 "We prefer to do things comfortably": Aldous Huxley, *Brave New World* (1932; Harper, 2017), 240.

INDEX